PRENTICE-HALL

FOUNDATIONS OF DEVELOPMENTAL BIOLOGY SERIES

Clement L. Markert, Editor

Volumes published or in preparation:

FERTILIZATION   C. R. Austin

CONTROL MECHANISMS IN PLANT DEVELOPMENT
Arthur W. Galston and Peter J. Davies

PRINCIPLES OF MAMMALIAN AGING
Robert R. Kohn

EMBRYONIC DIFFERENTIATION
H. E. Lehman

DEVELOPMENTAL GENETICS*   Clement L. Markert and
Heinrich Ursprung

CELL REPRODUCTION DURING DEVELOPMENT
David M. Prescott

CELLS INTO ORGANS: The Forces That Shape
the Embryo   J. P. Trinkaus

PATTERNS IN PLANT DEVELOPMENT   I. M. Sussex and
T. A. Steeves

*Published jointly in Prentice-Hall's *Foundations of Modern Genetics Series*

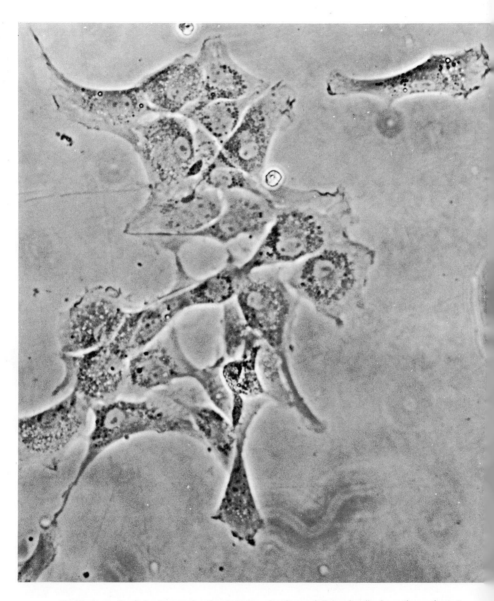

Frontispiece A photomicrograph of a primary clonal population of cells from the embryonic chick heart, growing on a plastic substratum. The cells in contact in the upper part of the colony do not overlap each other; hence they show contact inhibition. Some cells in the lower part of the colony, however, do overlap, and therefore apparently show less contact inhibition. Ruffled membranes may be seen at the flattened free edges of several cells of the colony. Note the absence of ruffled membranes where the cell surface is in contact with that of another cell. Ruffling is also evident at the broad leading edge of the single free cell. This cell has broken away from the cell cluster and is moving toward the right. Magnification X400. (Photograph by Linda S. Krulikowski.)

CELLS INTO ORGANS

The Forces That Shape the Embryo

J. P. Trinkaus

Yale University

PRENTICE-HALL, INC. Englewood Cliffs, New Jersey

© 1969 by PRENTICE-HALL, Inc.
Englewood Cliffs, New Jersey

All rights reserved. No part of this
book may be reproduced in any form
or by any means without permission
in writing from the publisher.

Printed in the United States of America
C— 13-121657-0
P— 13-121640-6
Library of Congress Catalog Card Number: 79-76877

Current printing (last digit):

10   9   8   7   6   5   4   3   2   1

PRENTICE-HALL INTERNATIONAL, INC., London
PRENTICE-HALL OF AUSTRALIA, PTY. LTD., Sydney
PRENTICE-HALL OF CANADA, LTD., Toronto
PRENTICE-HALL OF INDIA PRIVATE LTD., New Delhi
PRENTICE-HALL OF JAPAN, INC., Tokyo

For Madeleine

Foundations of Developmental Biology

The development of organisms is so wondrous and yet so common that it has compelled man's attention and aroused his curiosity from earliest times. But developmental processes have proved to be complex and difficult to understand, and little progress was made for hundreds of years. By the beginning of this century, increasingly skillful experimentation began to accelerate the slow advance in our understanding of development. Most important in recent years has been the rapid progress in the related disciplines of biochemistry and genetics—progress that has now made possible an experimental attack on developmental problems at the molecular level. Many old and intractable problems are taking on a fresh appeal, and a tense expectancy pervades the biological community. Rapid advances are surely imminent.

New insights into the structure and function of cells are moving the principal problems of developmental biology into the center of scientific attention, and increasing numbers of biologists are focusing their research efforts on these problems. Moreover, new tools and experimental designs are now available to help in their solution.

At this critical stage of scientific development a fresh assessment is needed. This series of books in developmental biology is designed to provide essential background material and then to examine the frontier where significant advances are occurring or expected. Each book is written by a leading investigator actively concerned with the problems and concepts he discusses. Students at intermediate and advanced levels of preparation and investigators in other areas of biology should find these books informative, stimulating, and useful. Collectively, they present an authoritative and penetrating analysis of the major problems and concepts of developmental biology, together with a critical appraisal of the experimental tools and designs that make developmental biology so exciting and challenging today.

Clement L. Markert

Preface

> And the end of all our exploring
> Will be to arrive where we started
> And know the place for the first time.
>
> T. S. Eliot, *Little Gidding*

This book is concerned with the origins and changes in form of tissues and organs. Although recognition of the importance of these phenomena is as old as embryology itself, sustained investigation of them is a modern development. Doubtless a major reason for this avoidance of problems so close to the heart of embryology was what Professor Ross G. Harrison liked to call the "gold rush" toward the analysis of embryonic induction, which for so long diverted attention from other problems of development. But another reason surely was the rather primitive state of our knowledge of cell surface behavior. In spite of pioneering studies by Harrison, Holtfreter, Lewis, Weiss, and others through the years, detailed knowledge of how cells contact each other, how they move about, and how their individual activities are converted into the collective movement of cell sheets was until recently largely lacking. Now, however, this information is becoming available, due in substantial degree to improved methods of culturing cells and the invention of a number of techniques that permit closer study of cell contact phenomena. For some time the study of normal tissue cells and cancer cells in culture and the study of mechanisms of morphogenetic cell movements in the embryo were largely separate, in detriment particularly to the latter. Now, however, the fields have converged. This is of course to their mutual profit, and progress recently has perceptibly accelerated, so much so in fact that study of the cell surface in all of its aspects is rapidly becoming one of the most active areas of cell and developmental biology.

    The purpose of this book is to try to pull together some of this work on cell contacts and movements, particularly as they relate to changes in the shape of multicellular systems during development. This is where my own interests lie, and it is in this area and in the area of cancer cell movement that I believe some of the most important problems ultimately lie. My intent has been to describe the current situation as it appears to me and to raise questions, rather than survey the field and record all that has been done. Since I have not striven for completeness, much has been left out or given limited consideration. Rather less attention, for example, has been devoted to what might be termed the molecular features of cell contacts than might be thought desirable. This is in part because at present this is an almost entirely speculative area with little substantial that has been really well-established and in part because the subject has been reviewed recently in depth in books by A. S. G. Curtis and L. Weiss. It is true, of course, that speculations of this sort can be useful, in that they may suggest what to look for in the laboratory. But it is also true that oftentimes questions must first be answered on a cell and fine-structural level before we are in a position to pose biologically meaningful questions on a physicochemical level. In any case, my abridged treatment of the subject certainly does not reflect conviction on my part that the solutions to problems of cell movements lie elsewhere. On the contrary, along with everyone else I believe that it is precisely in the molecular structure of the cell membrane, in the dependence of this structure on the genome or indeed in its independence of the genome, and in its relations to various substrata during cell contact and movement that solutions to the problems posed by tissue and organ shape changes are ultimately to be found.

Another area that has received relatively little attention in this book is plant morphogenesis. Much less space is given to plant cells than to animal cells. There is good reason for this. The higher plants do not utilize cell movements as a morphogenetic mechanism (and in this lies what is probably the most important difference between plant and animal morphogenesis). In certain of the lower plants, however, in particular the cellular slime molds, cell movements have supreme significance in morphogenesis. I have therefore dealt with them at some length.

The book is thus more an essay than a survey—an essay on some of the major problems that confront cell and developmental biologists when they think about cell contacts, cell movements, and morphogenesis. Hopefully, it will alert the reader to some of the areas that most desperately need attention and stimulate research activity. The book will have succeeded if it accelerates its already burgeoning obsolescence.

In keeping with the decision not to seek completeness, no consistent effort has been made to give detailed credit for work discussed. For this I ask the indulgence of my colleagues. However, since the purpose of the book is to stimulate interest in morphogenetic movements and cell contact phenomena, a means must be provided for moving beyond the text and finding the original works on which the discussions are based. To this end, a list of "selected references" is appended to each chapter. These titles have been selected partly because they describe work that I consider important and partly because they are recent and contain useful bibliographies. All investigations described in the text can be traced readily, either directly or indirectly, by means of these reference lists. A few works of a more comprehensive nature have been listed at the beginning to introduce the reader to subjects not covered or given only nodding attention in the present work. In a word, even though the reference lists are by no means complete, they can be used as an entry into the literature and a guide as to where to find it.

This volume was written in a small village in the south of France, where distracted only by the sun, the sea, and an occasional game of pétanque I was free to read, reflect, and write at length on problems close to my heart. I am indebted to my wife who gave up untold hours in the sun, with hardly a complaint, to act as my typist. I am also grateful to my colleagues Adam Curtis, Clement Markert, and James Weston, all of whom read the manuscript and made many suggestions, some of which I adopted. I am indebted to Linda Krulikowski and Dolores Slaughter for aid with the references, the figures, and the index. Finally, I wish to express my gratitude to students and colleagues at Yale University and at Woods Hole for many profitable discussions of these problems.

My research through the years and the writing of this book have been supported by grants from the National Science Foundation.

Hurlevent
Le Castellet (Var), France
August, 1968

J. P. TRINKAUS

Contents

General references xvi

ONE • INTRODUCTION: MORPHOGENETIC CELL MOVEMENTS   1

Gastrulation and neurulation   2
Movements of individual cells during embryogenesis   5
   Primordial germ cells 5
   The neural crest 7
   Blastulation and gastrulation 11
Individual cell movements in adults   13

TWO • MECHANISMS OF CELL LOCOMOTION   17

Cell movements in vitro   18
   Amoeboid movement 18
   Gliding movement 19
Contact inhibition   22
Adhesion to the substratum   26
Contact guidance   28
Cell movements in vivo   31

THREE • DIRECTIONAL MOVEMENTS AND CHEMOTAXIS   36

Trap action   37
Chemotaxis   37
   Bracken fern spermatozoa 38
   Animal cells 40
   The cellular slime molds 45

FOUR • THE STRUCTURAL BASIS OF CELL ADHESION   55

Fine structure of cell contacts   55
Cytoplasmic bridges   63
Cell fusion   64

FIVE • THE MEASUREMENT OF CELL ADHESION   68

SIX • MECHANISMS OF CELL ADHESION   76

Molecular complementarity hypothesis   76
Calcium-bridge hypothesis   79
Coordination about potassium ions   81
Adhesion of cells to glass   82
Physical theories   83
Isolation of cell membranes   88
Intercellular materials   90
Selective deadhesion   97

SEVEN • CELL SEGREGATION AND SELECTIVE ADHESION   100

Reconstitution of cellular aggregates   100
Theories of cellular segregation   109
    Chemotaxis   *109*
    Timing hypothesis   *110*
    Differential cellular adhesiveness   *110*
A critique of the theories of cellular segregation   112

EIGHT • MOVEMENTS OF CELL SHEETS   126

The spreading of epithelial sheets   126

NINE • SOME GUIDING PRINCIPLES IN THE STUDY OF MORPHOGENETIC MOVEMENTS   132

TEN · ECHINODERM INVAGINATION   **136**

ELEVEN · AMPHIBIAN GASTRULATION   **146**

TWELVE · NEURULATION   **159**

THIRTEEN · EARLY CHICK MORPHOGENETIC
            MOVEMENTS   **166**

    Epiboly of the area opaca   166
    Endoderm and mesoderm formation   169
    Migration of precardiac mesoderm   175

FOURTEEN · TELEOST EPIBOLY   **179**

FIFTEEN · EPILOGUE   **203**

    Principles of cell motility   203
    Microfilaments and microtubules   204
    Contact inhibition   207
    The cell membrane as an autonomous unit   210
    Genic control of morphogenetic movements   213

AUTHOR INDEX   **217**

SUBJECT INDEX   **221**

General references

Allen, R. D. and N. Kamiya. [ed.] 1964. Primitive motile systems. Academic Press, New York. 642 p.

Bonner, J. T. 1952. Morphogenesis. An essay on development. Princeton Univ. Press, Princeton. 296 p.

Curtis, A. S. G. 1967. The cell surface: its molecular role in morphogenesis. Logos Press and Academic Press, London and New York. 405 p.

Fawcett D. W. 1966. An atlas of fine structure. The cell, its organelles and inclusions. Saunders, Philadelphia. 448 p.

Gustafson, T. and L. Wolpert. 1967. Cellular movement and contact in sea urchin morphogenesis. Biol. Rev. Cambridge Phil. Soc. **42**:442–498.

Picken, L. 1960. The organization of cells and other organisms. Oxford, London. 629 p.

Trinkaus, J. P. 1965. Mechanisms of morphogenetic movements, p. 55–104. *In* R. L. DeHaan and H. Ursprung [ed.] Organogenesis. Holt, Rinehart & Winston, New York.

Trinkaus, J. P. 1966. Morphogenetic cell movements. p. 125–176. *In* M. Locke [ed.] Major problems in developmental biology. The twenty-fifth symposium of the Society for Developmental Biology. Academic Press, New York.

Waddington, C. H. 1956. Principles of embryology. Allen and Unwin, London. 510 p.

Weiss, L. 1967. The cell periphery, metastasis and other contact phenomena. North-Holland, Amsterdam. 388 p.

Weiss, P. 1961. Guiding principles in cell locomotion and aggregation, Exptl. Cell Res. Suppl. **8**:260–281.

Wilt, F. H. and N. K. Wessells. 1967. Methods in developmental biology. Crowell, New York. 813 p.

ONE

Introduction: Morphogenetic Cell Movements

The development of multicellular organisms has three fundamental aspects: the differentiation of cells as individuals, termed cytodifferentiation; the collective behavior of cells in the formation of tissues and organs, termed histogenesis or organogenesis; and growth or increase in mass. The study of cytodifferentiation requires research primarily at the subcellular level: how the genetic information encoded in the genes is transcribed and translated into the cytoplasmic end products that characterize each particular cell line, how this process is controlled by other cells during embryonic induction, and how differentiation in a particular direction becomes stabilized. The investigation of various aspects of cytodifferentiation is of course basic, and it is only by such research that we will eventually achieve understanding of how the embryo gives rise to so many different types of cells. Investigations of this sort, however, are not likely to lead to an understanding of how the different cells of an embryo become arranged in highly ordered multicellular systems.

To learn how a collection of various kinds of cells move about and make the specific and enduring contacts that create organs or, in a word, how one moves from the cellular level to the level of the organism requires study primarily at the cellular level. The specific objects of such study are the motile activities of cells and the adhesive contacts their surfaces make with one another. Certainly, these surface properties of cells also depend on the activities of genes and other subcellular constituents. Hence, as we learn in detail how cells move and make contacts the hard distinction we have drawn between the two levels of organization will become blurred and lose much of its utility. For the present, however, this distinction has merit. It calls attention to a vast area of cellular activity peculiar to multicellular organisms that until recently has been largely neglected and needs intensive study before connections at the genetic level can be studied with profit. Before we study the activities of genes we must know the phenotypes they produce.

The primary purpose of this book is to consider the properties of cells as they relate to normal morphogenetic movements and cell contacts during development. It is by means of morphogenetic movements that cells become arranged in the tissue patterns characteristic of each species. Such movements may involve translocation of cells over considerable distances to different parts of the embryo, as during the invagination of an amphibian egg or the migration of germ cells and cells of the neural crest. They may also involve an in situ change in form of a discrete structure, such as the indentation of an optic vesicle to form an optic cup or the expansion of the gut to form pharyngeal pouches. Movements of individual cells have undoubted importance in development, but these are not the dominant movements. It is by the spreading and folding of cohesive sheets of cells that most morphogenetic movements are accomplished. The most spectacular of these are of course gastrulation and neurulation, and it is these that have been most closely studied. As a consequence, in the vertebrates at least, the broad outlines of these movements are quite well known (although there is still a good deal of smaller-scale movement to be charted).

Gastrulation and neurulation

By means of gastrulation the hundreds or thousands of cells of the blastula translocate in defined and often dramatic ways, so that at the end of the process the various primordia become arranged in accordance with the primitive body plan of the organism. Ectoderm comes to lie at the surface, chordamesoderm beneath it, and the endoderm inside or lowermost. As a result of these rearrangements the embryo acquires a clear anteroposterior polarity and bilateral symmetry. Moreover, gastrulation is a phenomenon of the whole egg, involving all or almost all its cells. From various points of view, therefore, gastrulation is of fundamental importance in early animal development. An introductory text of embryology should be consulted for a detailed understanding of the normal course of gastrulation movements in some of their variety. Only certain salient features will be pointed out here.

In a holoblastic egg, such as that of the sea urchin, gastrulation appears to be a relatively simple process. There is little or no epiboly or spreading of the ectodermal blastocoel wall. Moreover, invagination is confined to cells of the vegetal pole, the so-called vegetal plate (Fig. 10-1). There is little yolk, hence invagination is unimpeded in these eggs and results in the formation of a long symmetrical endodermal archenteron or primitive gut. Mesoderm forms subsequently by the evagination of coelomic pouches from the wall of the archenteron.

It is in the amphibian egg that gastrulation has been studied most intensively. Indeed, the masterful investigations of Vogt on amphibian gastrulation in the 1920s were so complete that they stand as a monument to the possibilities of descriptive embryology. The presence of large amounts of yolk within the cells, particularly those of the vegetal half of the egg, modifies both cleavage and gastrulation. The vegetal blastomeres tend to be rather large and bulky, and the archenteron appears largely in the animal half of the egg, resting on the less active mass of vegetal cells. Archenteron formation is caused by invagination of cells at the blastopore and involution of cells over the lips of the blastopore. This involution of surface cells is sustained by a massive epibolic spreading of the remaining surface cells of the animal hemisphere to eventually cover the entire egg. As the prospective chordamesoderm and endoderm undergo involution to form respectively the roof and floor of the archenteron, they are replaced by the spreading prospective ectoderm. There is also strong convergence of cells undergoing involution in order that they may turn into the relatively small blastopore. Inside the egg, the walls of the archenteron spread laterally and anteroposteriorly. Thus, in amphibian gastrulation, except for some of the yolky vegetal cells, the entire egg is involved in extensive morphogenetic movements. Just after the completion of gastrulation, with closure of the blastopore, the ectoderm of the neural plate forms the neural folds, which roll together in another extensive movement to form the neural tube (Fig. 12-1).

In eggs such as those of teleost fishes, reptiles, and birds, the amount of yolk is so massive that it does not engage in cleavage, the latter being confined to cytoplasm located superficially at the animal pole. Cleavage is therefore meroblastic and results in a more or less flat cap of cells called a blastoderm, surmounting the relatively enormous yolk mass. Gastrulation occurs in a rather different fashion in teleosts as compared to reptiles and birds. Mammals gastrulate like reptiles and birds, in spite of their relative lack of yolk and holoblastic cleavage, presumably because of their evolutionary origins in the reptiles.

The teleost blastoderm begins gastrulation by flattening and spreading over the yolk in a spectacular movement of epiboly, which ultimately entirely encompasses the large spherical yolk mass. Soon after epiboly begins, the marginal region of the blastoderm thickens, forming the so-called germ ring (Fig. 14-1). The cells of this thickened region undergo strong dorsad movements that cause them to converge in the dorsal part of the blastoderm, forming the broad, thickened embryonic shield out of which the embryo proper forms. As epiboly continues, the germ ring thins, and the ever growing embryonic shield elongates and divides into germ layers and primitive primordia. Teleost gastrulation is clearly very different from that of the echinoderms or the Amphibia in that there is neither invagination nor involution. Moreover, the central nervous system forms by secondary hollowing out of a solid rod rather than by the rolling together of raised neural folds.

Chick gastrulation is equally meroblastic and begins by delamination of the hypoblast, which later forms the endoderm. The next processes are similar to the teleosts: an epibolic spreading of the entire blastoderm and a convergence of certain cells of the central pellucid area of the blastoderm toward the midline. But the resemblance ceases there. As surface cells of the epiblast reach the midline they sink down to form the elongate primitive streak and then spread out between the converging epiblast above and the hypoblast below to form the mesoderm. This convergence is most impressive. The entire posterior half of the epiblast sheet slides to the midline to fold into the primitive streak. In this manner, cells as far distant as the margin of the pellucid area eventually come to invade the streak. Convergence of the epiblast and streak formation (and of course involution of cells at the primitive streak) continue until the streak has elongated considerably. By then, much mesoderm has formed, and chordamesodermal structures soon become evident anterior to the streak. These are a prelude to embryo formation. As new chordamesoderm cells are added, the embryo elongates. As the embryo elongates, the primitive streak regresses, until, finally, with the completion of embryo formation, the streak is incorporated into the tail bud. While all this is occurring, the blastoderm expands continually over the massive yolk, a process that continues long after the completion of gastrulation and is essential for the formation of the yolk sac.

There are of course other cell movements, besides those of gastrulation, that are essential for morphogenesis. Some of these have been studied in detail, and, like gastrulation, they involve mainly (but by no means exclusively) the spreading and folding of cohesive sheets of cells. The brain expands to form its various divisions, including the optic vesicles. These invaginate and form the optic cups. Ectoderm in contact with the optic cup invaginates to form the lens vesicle. The olfactory and auditory vesicles form in a similar fashion. But such movements are not confined to the ectoderm. The mesodermal ureteric bud and nephrogenic blastema expand and join to form the convoluted nephron. The gut and its derivatives achieve their final form as a result of manifold evaginations, foldings, and extensions. Perusal of any standard text on developmental anatomy will reveal other examples. Even though the exact amount of spreading of the cell sheet in each of these is generally not known, because of their inaccessibility, there is little doubt that much spreading occurs. In addition, in both embryos and adults various epithelia respond readily to wounding, spreading over surfaces of considerable area.

Movements of individual cells during embryogenesis

Although the spreading and folding of cohesive cell sheets are the primary means of rearranging cells during development of the Metazoa, it is not the only mode of cellular locomotion. Some cells move as individuals, wending their way through the interstices of the embryo to accumulate, often at considerable distances from their point of origin, at new loci where they participate in histogenesis. Well known examples are the primordial germ cells and cells of the neural crest.

Primordial germ cells

These cells move relatively long distances during early embryogenesis to settle finally in the germinal epithelium. As the latter develops into a gonad, the primordial germ cells are incorporated in the differentiating seminiferous tubules and ovarian follicles and become spermatogonia or oogonia. In the chick embryo, primordial germ cells are first detected as large cells in the "germinal crescent" in the extraembryonic endoderm anterior to the embryo of a head process stage, whence they travel posteriorward, eventually to reach the gonadal area. If a blastoderm whose germinal crescent has been excised is joined in culture with a normal blastoderm in such a fashion that yolk-sac blood vessels anastomose, the gonads of the former will be colonized by the germ cells of the latter (Fig. 1-1). With this remarkable technique Simon was able to form various chimeric gonads. Chick gonads, for example, have been populated by the germ cells of a duck.

Primordial germ cells apparently migrate best through blood vessels, for in the absence of circulatory anastomoses the gonads of an embryo lacking a germinal crescent remain sterile. But just how might they move in the blood vessels? Are they carried passively, or do they migrate actively,

6   Introduction: Morphogenetic Cell Movements

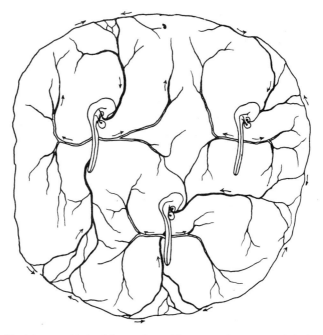

Fig. 1-1   Circulatory parabiosis of three embryos. Note a vascular anastomosis in the center. The two anterior embryos communicate directly by way of a vitelline vein with the posterior embryo. (Simon. 1960. Arch. Anat. Microscop. Morphol. Exptl. **49**:93.)

with the walls of the blood vessels serving as guides? These questions can be settled only by following the cells. Meyer identified primordial germ cells by differential cytochemical staining. These cells have high glycogen content which can be stained specifically with PAS (Periotic acid–Schiff reagent). Germ cells first appear within the embryo at the onset of cardiac propulsion and blood circulation. It seems, therefore, that they are wafted passively in the circulation. Initially, they are found throughout the circulatory system, but by stages the majority becomes concentrated in the future gonadal regions, where they accumulate in small blood vessels leading from the dorsal aorta. It is not clear why they accumulate there. It could be because of lack of passage due to the small size of the vessels or because of more subtle factors, such as differential adhesiveness of their cell surfaces for the cells of this region.

Even though there can be no reasonable doubt that germ cells of the embryonic gonad of the chick derive from the germinal crescent, there is still no evidence that they are the source of the sperm and eggs of the adult. This is a difficult question to settle. It is therefore with particular interest that we turn to a recent study of germ cells in the mouse, which has provided direct proof that primordial germ cells originating in the

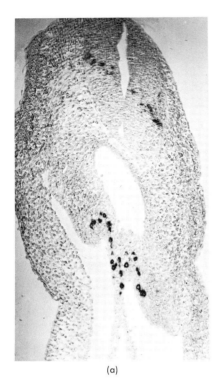
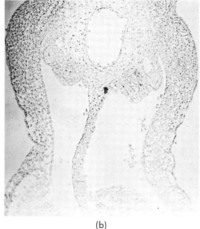

Fig. 1-2 (a) Normal 10-day mouse embryo. Primordial germ cells stained for alkaline phosphatase may be seen entering gonadal folds from dorsal gut mesentery via coelomic angles. ×115. (b) Mutant (WW^V) littermate of embryo in Fig. 1-2a, with one germ cell near the base of the mesentery. ×115. (Courtesy of B. Mintz. Mintz. 1959. Arch. Anat. Microscop. Morphol. Exptl. **48**: Suppl. 155.)

(a) (b)

endoderm not only migrate to the mesodermal gonads but once there persist and eventually give rise to sperm and eggs in the adult. Mintz observed that primordial germ cells of the mouse have high alkaline phosphatase activity and so can be readily identified cytochemically and counted while still in the early endoderm. In a mutant strain, which is almost always sterile, the number of primordial germ cells in the yolk sac endoderm of the embryo is correspondingly reduced (Fig. 1-2).

The neural crest

Cells of the neural crest migrate from their origin on the dorsal surface of the neural tube to various distant areas of the embryo and form products as diverse as sympathetic ganglia, pigment cells, spinal ganglia, adrenal medulla, and visceral cartilages. Because of the importance and dispersion of its products, this transitory embryonic primordium has attracted the attention of many investigators during the last half century. During this period the list of histological end products has lengthened to the point where the neural crest stands unchallenged among embryonic primordia for the diversity of its derivatives. Yet many questions remained unanswered. Do all cells of the neural crest find their mark? Or do many of them disperse into other areas, where they either cytolyze, remain undifferentiated, or are incorporated in undetectably small numbers in the structures which form in that area? Do cells of the neural crest migrate at random from the neural tube out into the embryo, or do they move along certain pathways? If crest cells move along certain pathways, are these characteristic and well defined? If so, what features of the cells and their environment serve to direct their movements?

None of these questions could be answered until a cell-specific, nontoxic, long-term cell marker had been developed. The classical method of following sheets of cells with vital dyes is inadequate for tracing individual cells, because the dye is not permanent and may transfer to other cells. Surface markers such as carbon particles do not work because they are rubbed off as the cells push between other cells. Nuclear size markers have been useful, but because the range of nuclear sizes overlaps in donor and host this method is not always reliable. A marker with the appropriate qualifications finally became available with the synthesis of *tritiated thymidine*. This nucleoside is permanently incorporated into DNA in the nucleus prior to each cell division, is not deleterious at moderate dose levels, and gives high resolution radioautographs as a result of the low-energy beta emanations from tritium. Thus, tritiated thymidine can give cell-to-cell labeling wherever DNA synthesis occurs, regardless of tissue, stage, or organism (Fig. 1-3). This is obviously an ideal means of tagging neural crest cells. When marked neural crest cells are grafted to an unlabeled host, the movements of the labeled cells can be followed. Weston has done this recently in the chick embryo and has provided answers to some of the questions posed above.

Cells of the trunk neural crest emigrate in two rather well-defined streams of individual cells—one leading ventrad into the mesenchyme between the neural tube and the myotome, the other dorsad into the superficial ectoderm. The migration is in no sense random. The cells obviously follow favored pathways and then, for reasons not yet understood, cease their movement and accumulate rapidly in known points of destination, such as sensory ganglia and sympathetic ganglia (Fig. 1-4). The dorsad stream moves immediately into the ectoderm. When ectoderm containing these cells is grafted to a nonpigmented host it yields pigment cells, supporting the suggestion that these cells are promelanoblasts. This is of some interest and illustrates nicely the usefulness of a direct method. It had previously been thought on the basis of indirect evidence that promelanoblasts were confined to the dermis during their migration from the crest.

Interestingly enough, the direction of the ventrad migration through somitic mesenchyme is independent of the mesenchyme through which

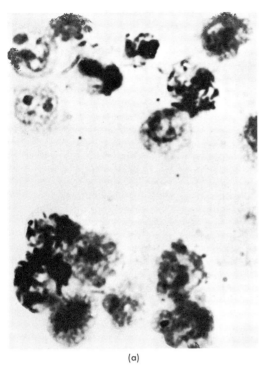

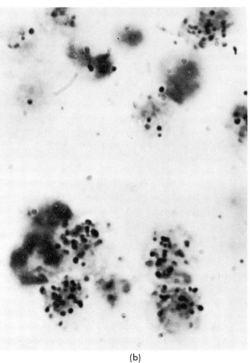

Fig. 1-3 (a) A reaggregate of labeled 4-day mesonephric cells and nonlabeled 6-day retinal pigment cells, prior to culturing. Note five pigment cells. ×100. (b) Autoradiograph of 1-3a. Radioactivity is confined to mesonephric cells. The apparent low activity of the pigment cell at upper left is probably a chance concentration of mechanically activated silver grains. ×1800. (Trinkaus and Gross. 1961. Exptl. Cell Res. **24**:52.)

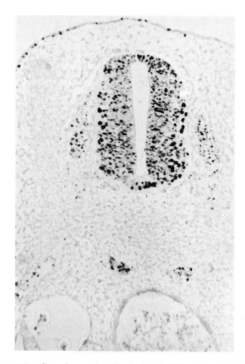

Fig. 1-4 Transverse section through seventeenth somite of chick embryo fixed at 4 days of incubation. A piece of neural tube (with neural crest) labeled with tritiated thymidine had been grafted to this embryo 50 hours previously. Radioautograph exposed 14 weeks. Note labeled cells in ectoderm, condensed labeled spinal ganglia, sheath cells on motor nerve in lower left corner, and labeled sympathetic ganglia adjacent to aorta (bottom). ×150. (Courtesy of Jim Weston. Weston. 1963. Develop. Biol. 6:279.)

it moves. It is somehow related to the orientation of the neural tube. Thus, when a neural tube with labeled neural crest is grafted in an inverted orientation, neural crest cells continue to migrate ventrad relative to the neural tube, which of course is now dorsad relative to the host embryo. Metamerism of sensory and sympathetic ganglia, on the other hand, is strongly influenced by the mesodermal somites. Early migration is for some reason enhanced within the segmented somitic mesenchyme, whereas migration between the somites is attenuated. The labeled cells that migrate first through the somitic mesenchyme localize at the dorsolateral margin of the aorta very early in development. These presumably become sympathetic neuroblasts.

The use of tritiated thymidine has thus solved the problem of random migration versus migration along particular pathways in favor of the latter. It has also provided impressive confirmation of the neural crest origin of certain derivatives. The cells apparently migrate to the loci where their derivatives will form and go nowhere else. This is testimony not only to the reliability of tritiated thymidine as a long-term cell marker but also to the care and accuracy with which early workers drew their conclusions from the circumstantial evidence provided by grafting and deletion experiments. One problem of the neural crest, however, was not solved by the old methods—the origin of the Schwann sheath cells. The question of whether they are derived solely from the crest, as maintained by some, or are in part derived from the neural tube, as contended by others, remained unresolved until the tritium label was applied. Labeled neural tube from which the crest has been removed gives rise to labeled sheath cells when grafted in situ. The controversy is thus in part settled. Some (perhaps all) Schwann sheath cells are derived from the ventral neural tube in the chick embryo. Incidentally, some of these cells must migrate spectacular distances in the embryo in order to adhere eventually to peripheral nerve bundles.

Blastulation and gastrulation

The blastula had been considered in the past to be merely the prelude to gastrulation and not a stage during which cell movements occur. Some years ago, however, Schechtman and Nicholas found that when they stained the vegetal blastomeres of an amphibian blastula with vital dye, much of the dye soon disappeared from the surface. Dissection revealed that some inner cells were now stained. This exciting discovery was at first interpreted as evidence that individual blastomeres at the vegetal pole slip inside in a movement of "unipolar ingression." If true, a new morphogenetic movement had been found. However, a subsequent investigation by Ballard, in which cell lineage at the vegetal pole was meticulously followed, showed that no cells leave the surface. Stain is carried inside by the expanding cortex or surface layer of the large vegetal blastomeres as they divide. Although this is in itself an interesting and undoubtedly important phenomenon in view of the morphogenetic importance of the cortex, ingression of whole individual cells during the blastula stage was disproved.

At the onset of gastrulation, on the other hand, individual cell movements are very much in evidence. Elongate "bottle" cells move individually into the interior of an amphibian egg at the site of the future blastopore, retaining only a stretched, tenuous connection with the surface. They presage the invagination of the prospective endoderm and mesoderm and mark the beginning of gastrulation (see p. 147).

Movements of individual cells are also important in sea urchin gastrulation. As in the amphibian egg, cells leave the surface to move inside precisely at the point where invagination will occur. This inner movement

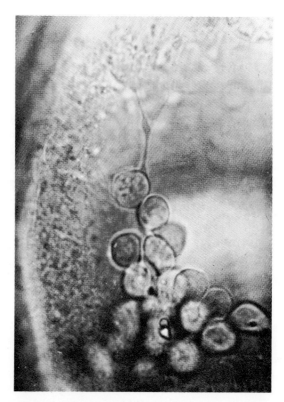

Fig. 1-5 Pseudopodial attachments of primary mesenchyme cells to the ectoderm in early development of the sea urchin *Psammechinus miliaris;* a cable-like exploring pseudopod extends in the direction of the animal region. Magnification about × 1000. (Courtesy of T. Gustafson. Gustafson and Wolpert. 1961. Exptl. Cell Res. **24**:64.)

of cells from the vegetal plate into the blastocoel, where they later form the primary mesenchyme cells, has been known for decades and can be observed easily in the transparent living blastulae (Fig. 1-5). Later on, the emergence of a secondary group of mesenchyme cells from the tip of the advancing archenteron is likewise easily observed. These cells connect the archenteron wall to the blastocoel wall by means of long filopodia and by this activity probably have great importance in the mechanism of invagination. The transparency of teleost eggs also permits observation with time-lapse cinemicrography, and it has been shown that here too cells move extensively as individuals during gastrulation. In this case, it is the so-called "deep" cells or inner cells of the blastoderm that migrate. Their migrations account in part for epiboly and embryo formation.

The significance of amphibian bottle cells, sea urchin mesenchyme, and teleost deep cells for gastrulation are subjects that will receive more attention later. Movements of cells as individuals have been suspected to be an important feature of gastrulation in many forms. But nowhere outside the amphibia, sea urchins, and teleosts have their migratory power and relevance for gastrulation been clearly demonstrated.

Individual cell movements in adults

Movements of individual cells are not confined to embryos. Amoebocytes of various kinds have extraordinary migratory powers in both embryos and adults and can move to and accumulate rapidly in great numbers at a focus of infection. The activities of lymphocytes and other leucocytes in this regard are well-known. They are carried mainly in the blood and lymph, in which they may be transported passively to all parts of the organism. In addition, they can squeeze out of these vessels and by active movements propel themselves through tissues. Recent evidence suggests that lymphocytes populate the spleen and lymph nodes, their centers of activity and reproduction in the adult, after migration from the distant thymus gland during late embryogenesis. If confirmed, this would constitute a morphogenetic movement on the part of individual cells that surpasses the distances traveled by cells of the neural crest and resembles that of the primordial germ cells, since they both utilize the vascular route.

Melanoblasts of adult birds retain the remarkable migratory powers of their distant neural crest ancestors. If barb ridge epidermis of an adult feather germ containing pigment cells is grafted to the base of the wing bud of an embryo, the donor melanoblasts may populate feather germs of the entire wing and much of the adjacent back and breast as well (Fig. 1-6). They will also migrate through the skin tissues of a young adult to invade a melanoblast-free piece of skin that is grafted after hatching.

The existence in the adult of pleuripotent migratory cells that are able to differentiate into a number of different tissues has long been postulated as a means of accounting for regeneration and cancer. These hypothetical cells are usually called *neoblasts* and have as one of their remarkable characteristics the capacity to migrate long distances to the site of the future neoformation. An elegant experiment by Butler and O'Brien eliminated the possibility that neoblasts come from afar to form the amphibian regeneration blastema. If a salamander is X-irradiated, regeneration does not occur. In the experiment, the entire animal except for part of the forelimb was irradiated. The limb was then amputated through unirradiated tissue only 0.5 mm away from the irradiated tissue. Normal, complete limb regeneration occurred. Whether neoblasts are lurking in the critical 0.5 mm is not known. But even if they are, they contribute little

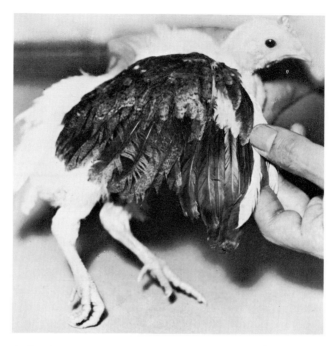

Fig. 1-6 A white leghorn chicken, which as a 3-day embryo received a graft from a feather germ of an adult brown leghorn. The melanoblasts from the adult donor migrated to the tip of the developing wing and penetrated its forming feather germs, there to differentiate into melanocytes which deposited melanin pigment in the epidermal cells of the developing feathers. The pigmentation is characteristic of the brown leghorn donor. (Trinkaus. 1953. Proc. Third Conf. Pigment Cell Growth. Academic Press, New York. p. 73.)

to the regenerate. The blastema has been demonstrated over and again to be composed of cells largely or entirely derived from the dedifferentiated tissue cells of the stump.

The status of the neoblast problem is less clear among the invertebrates. Large cells with basophilic cytoplasm, rich in RNA, are found distributed throughout Planaria and for a long time have been thought to be totipotent neoblasts which migrate to the area of a wound and give rise to the regenerate. Because the cells presumed to be neoblasts are scattered about in the organism it has been very difficult to do critical experiments. DuBois and Wolff found that an X-irradiated Planaria will regenerate at its anterior end if its tail is replaced by a nonirradiated tail and interpreted this as support for head formation by neoblasts which have migrated from the tail. Flickinger, on the other hand, questions this conclusion. He grafted pieces of tissue labeled with $^{14}CO_2$ to the prepharyngeal region and then decapitated the animal anterior to the graft. Regeneration occurred normally, even though there was no migration of cells to the blastema from the graft. In fact, radioautographs showed no movement of label at all from the graft. This result is clearly not in accord with the claims for extensive migration of neoblast reserve cells. On the contrary, it supports the concept of a local origin of blastema cells. Thus the results on this important question are contradictory. It appears that until we have direct evidence for the migration of neoblasts we had best accept tentatively the concept of local origin from differentiated cells.

The translocation over long distances of neoplastic cancer cells to metastasize in other tissues and organs is at the same time one of the most spectacular examples of movement by individual cells and small cell clusters in adults and a major reason for interest in cell movements within the organism. Perusal of any pathology text reveals many examples of cancerous growths appearing in diverse loci, far from the organ in which they were initially detected. Apparently the lowered adhesiveness (see p. 27) of many malignant cells permits cells to slough off the invading tongues of a tumor, thence to be carried away in the circulation. To what extent individual tumor cells can migrate actively through tissues is still a moot question.

Whether normal mesenchyme cells and tissue fibroblasts shift their positions in vivo is unknown. Their looseness of organization in vivo and their high degree of motility in vitro is suggestive, but not sufficient evidence on this question. This subject has been investigated recently by joining a piece of liver whose cells were labeled with tritiated thymidine to another piece of liver whose cells were unlabeled. Even after several days in culture there was no evidence of invasion of one by cells of the other. This result suggests that cells in intact tissues remain quite stationary. On the other hand, Berman, in Wolff's laboratory in Paris, and we ourselves have shown that when intact pieces of *different* tissues are juxtaposed in organ culture there is mutual invasion. It is important to note, however, that the invasion is by tongues of cells from the tissue mass, not by individual cells. In a similar vein, Steinberg has shown that when different tissues are joined by a glass fiber and rotated in fluid culture, one tissue will spread over the other, but invasion by individual cells will not occur.

SELECTED REFERENCES

Balinsky, B. I. 1966. Textbook of embryology. Saunders, Philadelphia. 673 p.

Ballard, W. W. 1955. Cortical ingression during cleavage of amphibian eggs, studied by means of vital dyes. J. Exptl. Zool. **129**:77–98.

Bonner, J. T. 1950. Morphogenesis. Princeton Univ. Press, Princeton, N.J. 296 p.

Du Bois, F. 1949. Contribution à l'étude de la migration des cellules de régénération chez les planaires *Dulcicoles*. Bull. Biol. Franc. Belg. **18**:213-283.

Hörstadius, S. 1950. The neural crest. Oxford Univ. Press, Oxford. 111 p.

Jacoby, F. 1965. Macrophages, p. 1-93. *In* E. N. Willmer [ed.] Cells and tissues in culture, vol. 2. Academic Press, New York.

Meyer, D. B. 1964. The migration of primordial germ cells in the chick embryo. Develop. Biol. **10**:154-190.

Mintz, Beatrice. 1959. Continuity of the female germ line from embryo to adult. Arch. Anat. Microscop. Morphol. Exptl. **48**:155-172.

Schechtman, A. M. 1934. Unipolar ingression in *Triturus torosus:* a hitherto undescribed movement in the pregastrula stages of a urodele. Univ. Calif. (Berkeley) Publ. Zool. **39**:303-310

Simon, Doris. 1960. Contribution à l'étude de la circulation et du transport des gonocytes primaires dans les blastodermes d'oiseau cultivés *in vitro*. Arch. Anat. Microscop. Morphol. Exptl.: **49**:93-176.

Trinkaus, J. P. 1953. Estrogen, thyroid hormone and the differentiation of pigment cells in the brown leghorn. Proc. Third Conf. on the Biology of Normal and Atypical Pigment Cell Growth. Academic Press, New York. p. 73-91.

Vogt, W. 1929. Gestaltungsanalyse am Amphibienkeim mit örtlicher Vitalfärbung. II Teil. Gastrulation und Mesodermbildung bei Urodelen und Anuren. Arch. Entwicklungsmech. Organ. **120**:384-706.

Weiss, L. 1967. The cell periphery, metastasis and other contact phenomena. North-Holland, Amsterdam. 388 p.

Weston, J. A. 1963. A radioautographic analysis of the migration and localization of trunk neural cells in the chick. Develop. Biol. **6**:279-310.

Weston, J. A. 1967. Cell marking, p. 723-736. *In* F. H. Wilt and N. K. Wessells [ed.] Methods in developmental biology. Crowell, New York.

Willis, R. A. 1952. The spread of tumors in the human body. Butterworth, London. 447 p.

TWO

Mechanisms of Cell Locomotion

Experimental study of the mechanism of morphogenetic movements is a completely modern subject which had its beginning a mere twenty-odd years ago. This is not to say that the mechanism of these cell movements did not concern embryologists during the early years of this century or indeed during the 19th century. As they gradually became aware of the extensive and spectacular nature of the cell movements of gastrulation from their studies of whole living eggs and sections of fixed embryos, they found themselves seeking explanations. The unavailability of techniques for an analytical approach did not prevent some provocative speculating. One of the earliest ideas was that a high rate of cell division in the animal hemisphere of the amphibian egg would cause the outer cell layer to spread and push less active cells into the interior, thus accounting for both epibolic spreading and invagination. The resorption of blastocoelic fluid that occurs coincidentally with invagination in the amphibian egg was thought by others to exert a sucking effect on the outer cell layer, causing it to invaginate. Still others postulated that a decrease in surface tension of the inner surface of cells of the vegetal half of the egg would cause that surface to expand and thus cause an inpocketing. The wedge or bottle shape of the first cells to invaginate is consistent with this idea. However such changes in cell shape could also be due to differential water inhibition, also a favorite idea for some time.

Some of these ideas were clearly ingenious and worthy of their creators, all of whom were leaders in the study of development. Nevertheless, since by and large they were not yet testable on the living embryo, they gave rise to no systematic research effort. The inventors were restrained both by lack of techniques for analysis and detailed knowledge of the normal course of morphogenetic movements.

Adequate descriptions of gastrulation did not come until the 1920s, with Vogt's studies on amphibian eggs. These were followed by, and indeed served as models for, similar studies of gastrulation of the chick and teleost and of other vertebrate types. With the direction, extent, and timing of the movements thus laid out, the stage was set for the only kind of approach that would yield explanations—analysis of individual cell movements, separately and in relation to each other.

Such an approach has impelled the study of cell movements and cell contacts largely in the artificial circumstances of cell and tissue culture, where in general the detailed behavior of cells can be observed and controlled far better than within the organism. It is not yet clear how much of what we learn from cells in vitro can be applied in vivo. At present, however, we have little choice but to learn as much as we can by this means, with at least the hope that from such studies will emerge fruitful ideas for solving problems of cell movements within the organism. In this and the next chapters, therefore, considerable attention will be devoted to the behavior of cells in culture.

Cell movements in vitro

It was once thought that only unicellular organisms and certain specialized tissue cells such as macrophages could migrate on solid substrata. With the invention of tissue and cell culture, however, it has been established that tissue cells in general can move readily over solid substrata under the proper conditions, whether this involves outgrowth on the substratum after escape from the periphery of an explanted chunk of tissue or directly from a dissociated suspension of cells.

Amoeboid movement

Most early observations of cell movement were made on free-living amoebae. Every student of elementary biology learns of the postulated roles of the "plasma sol" and "plasma gel" in amoeboid movement. In recent years there has been renewed interest in amoeboid movement, with the opportunities afforded by cinemicrography, viscosity measurements, micromanipulation, and the electron microscope. From these studies two dominant theories have emerged. The more classical point of view, recently promoted most actively by Goldacre, contends that forward movement results from a contraction of gelated proteins in the tail, causing solated endoplasm to flow forward toward the leading pseudopod. At the anterior end the endoplasm moves to the surface and gelates as ectoplasm which flows backward. As the gel moves posteriorward, it solates again and is again pushed forward. Thus Goldacre postulates gel membrane breakdown in the tail and continous reformation de novo in the anterior end. Two facts favor the posterior contraction theory. Microinjection of adenosine triphosphate (ATP) into the anterior end causes reversal of streaming. Microinjection of ATP into the tail causes an increase in the amoeba's forward speed. On the other hand, the most recent evidence is decisively against continual renewal of the membrane during movement. To test this hypothesis, Wolpert and O'Neill isolated surface membranes, prepared antibodies against them, labeled the antibodies with fluorescein, and reacted the labeled antibodies with living amoebae. The results showed no indication of surface renewal rapid enough to be consistent with the Goldacre postulate.

The major alternative theory is that of Allen. He contends that the driving mechanism is in the leading pseudopod itself, the contraction caused by the sharp turn of the forward flowing endoplasm at the pseudopod tip (the "fountain zone") providing the forward pull and anterior displacement of the endoplasm. A continuous forward stream is maintained by propagation of the contraction posteriorly along the axial endoplasm at the same rate as that at which the endoplasm advances. This theory requires that the flowing axial endoplasm be a weak-structured gel in order for the contraction to be propagated in it. Measurements show that the anterior endoplasm indeed has high viscosity. Accelerations required to effect particle displacement are higher for the anterior half of the animal, and amoebae subjected to centrifugal force accumulate particles on the centripetal side of the axial endoplasm. In addition, Allen has shown that portions of cytoplasm isolated in a capillary tube will develop an amazingly normal fountain streaming with an indication of contraction at the front.

Although the evidence is not yet conclusive for either theory, it appears at present to favor frontal pull rather than posterior push. Some workers believe different kinds of amoebae may utilize different mechanisms.

Whatever the motive forces may be, it seems possible that the same mechanisms are also at work in wandering amoeboid tissue cells, such as leucocytes and macrophages. Beginning investigations of these cells, however, do not provide a clear answer. Indeed, they suggest that there may be important differences. More intensive study of the movements of these cells is a pressing need and now possible with modern microscopy, microsurgery, and methods of cell culture.

Gliding movement

Other kinds of tissue cells, such as fibroblasts and epithelial cells, whose position in vivo is probably either fixed or changes little, will nevertheless move readily in culture and during wound healing, if provided with the proper substratum. They do not move by amoeboid movement. For these cells a solid or semisolid substratum is indispensable. They flatten and adhere firmly to the substratum and glide across it without pseudopodial formation or other gross deformation of cell form and without protoplasmic flow. Such cells can also move off the substratum across a fluid gap, but only if they can first attach extensions of themselves to a surface on the other side.

Certain plant cells also possess gliding movement. One type that has been the subject of intensive study is the blue-green alga *Oscillatoria*. Movement over the substratum is particularly spectacular in these cells, since they possess rather rigid membranes and often move as linearly arranged groups of cells, without apparent change in the shape of the individual cells. They truly give the impression that they are guided by an unseen force. The movement of these cells has recently been studied with a microscope specially adapted to reveal regions of contact of the cell with the substratum (surface-contact microscope). Such studies reveal that the movement of *Oscillatoria* and other algae is accompanied by delicate waves of contact between the cell membrane and the substratum, which travel continuously along the length of the filament (Fig. 2-1).

Waves of this sort could provide a mechanism for cell movement if they are accompanied by a compression wave also traveling along the length of the cell. If the compression wave is out of phase with the undulating wave, each succeeding contact with the substratum would take place at a different point on the membrane. Such a mechanism would provide forward motion in a manner rather like the gliding of an earthworm.

Fig. 2-1 (a) Undulating contacts between a filamentous alga and the solid substratum. C, contacts between membrane and substratum. (b) Mechanism of cellular locomotion involving undulating contacts between membrane and substratum. C and C', moving contact; T, transverse undulations of membrane; M, hypothetical compression wave traveling in cytoplasm. (After Abercrombie and Ambrose. 1962. Cancer Res. **22:**525.)

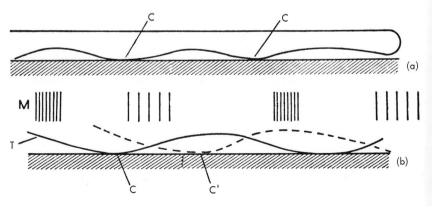

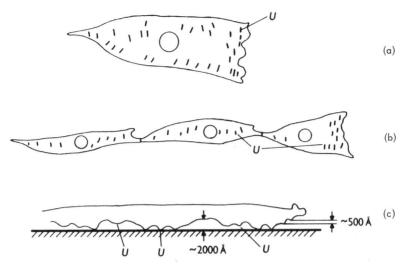

Fig. 2-2 Schematic drawings of fibroblasts in dark-field surface-contact microscope (lines are illuminated areas which appear to be in continuous undulating movement). (a) single fibroblast, (b) streaming group, (c) interpretation of image. U, undulating contacts. (Ambrose. 1961. Exptl. Cell Res. Suppl. 8:54.)

Although such a compression wave would appear to be a necessary accompaniment of the undulating wave in order for movement to occur, there is as yet no direct evidence for it.

Fibroblasts and epithelial cells appear to move in the same fashion, by waves of adhesive contact with the substratum. There is no evidence that protoplasmic flow is involved. And there is certainly no forward flow. Phase and interference microscopy have revealed that during movement the cell is attached only at its leading edge and its trailing edge. The leading edge of a moving fibroblast has the most attachments and consists of an exceedingly thin, fanlike membrane, 5–10 μ wide and closely applied to the substratum (Fig. 2-2). This membrane undergoes continual folding movements which beat inward (backward) from the edge of the membrane. These so-called ruffles occur rapidly enough to be seen without the aid of time-lapse cinemicrography. They are obviously stuck to the glass substratum, for they leave fine cytoplasmic filaments behind when they withdraw. Usually where a large ruffled membrane is seen the cell is moving in that direction, and when a new major ruffled membrane forms the cell soon begins to move in that direction.

From all these observations it appears that the ruffled membrane is the locomotory organ of the cell. Accompanying compression waves probably assure cell movement. There is no explanation as yet for wherein lies the "urge to move" that initiates formation of a ruffled membrane and consequent gliding movement.* Moreover, it is not sufficient for a cell to have a ruffled membrane in order to move. Isolated fibroblasts show little movement, even though they have pseudopods and ruffled membranes. Such cells show two or three opposed ruffled membranes and may fail to move because the locomotive forces are balanced in opposite directions. Weiss suggests that movement occurs when one or more ruffled membranes are suppressed, as in contact inhibition, so that as a result only one unopposed ruffled membrane is left.

Even though the leading edge of both fibroblasts and amoebae is active in directing movement of the cell, the mechanism is completely different in the two kinds of cells. In the fibroblast it is not at all accompanied by protoplasmic streaming toward the leading edge, as in the amoeba. Epithelial cells have been much less studied than fibroblasts, but it appears that they too move at the edge by a similar mechanism. Since most tissue cells are not amoeboid, it seems probable that the mechanism we have just described is the means by which most cells translocate. Incidentally, it appears that contractile proteins energized by ATP may be involved in the compression that doubtless accompanies the undulating waves. Glycerol-extracted fibroblasts contract in the presence of ATP in a manner reminiscent of skeletal muscle fibers.

A third class of cells that appears to move as a result of intermittent adhesive contacts with the substratum are the migratory, metastatic cancer cells. They differ from fibroblasts, however, in the number of flattened undulating fans. Fibroblasts tend to have one fan with an undulating membrane at the leading edge of the cell. Sarcoma cells (cancer cells of mesenchymal origin) tend to develop a number of such fanlike extensions. Very anaplastic cells (cancer cells that have lost all traces of tissue differentiation) are generally more rounded and the membrane may be divided into large numbers of small independently moving villi.

Contact inhibition

When a moving fibroblast contacts another fibroblast, it forms a massive contact with it, the cell contracts somewhat, the ruffled mem-

*Although the detailed relation of the ruffled membrane to cell movement is still unknown, independent studies by V. M. Ingram (in London) and by A. K. Harris (in our laboratory) have provided some information on the genesis of the ruffling. By observing fibroblasts cinematographically from the side, they have shown that a ruffle forms by the upward folding of the flattened cell margin (see Frontispiece). Once formed, a ruffle is propagated inward from the margin at a speed of about $\frac{1}{4}-\frac{1}{2}$ micron per second. It is not known what causes this uplift. Two possible explanations are decreased local adhesion of the edge to the substratum and contraction of the upper cytoplasm of the flattened edge.

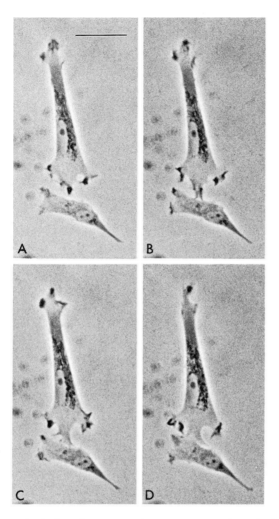

Fig. 2-3 Inhibition of ruffled membrane activity by contact with another cell. Photomicrographs of heart cells from a 9-day chick embryo, growing in culture on a plastic substratum. The photographs were taken from a time-lapse film. The scale equals 50 μ. (a) The upper larger cell, with a high degree of ruffled membrane activity at its leading edge, is approaching another cell. (b) Thirty-two seconds later. The middle part of the ruffled membrane has just contacted the other cell. Ruffling activity continues in the part in contact but appears diminished. (c) Fifty-six seconds later. Ruffled membrane activity has just ceased in the membrane immediately adjacent to the points of contact. Note the highly local nature of the inhibition. Ruffling continues in the rest of the leading edge on either side of the inhibited portion. No contraction of the cell is apparent. (d) Two minutes and 28 seconds later. The contact of the central part of the leading edge with the other cell has expanded. In spite of this and the length of time in contact, ruffling continues in the adjacent parts of the leading edge. (Trinkaus, Betchaku, and Krulikowski, unpublished.)

brane is paralyzed, and movement stops. Although such cells are not cemented rigidly together, their adhesion tends to persist about twenty minutes or even for several hours or longer. That this is primarily a surface reaction is indicated by the local nature of the inhibition and by the fact that if some surface is still exposed, a new ruffled membrane will soon develop there and the cell may break away again. In fact, movement begins again only when a new ruffled membrane forms (Fig. 2-3).

This tendency of fibroblasts to cease movement upon contacting other fibroblasts was discovered by Abercrombie and is called "contact inhibition of cell movement." It seems to be of great importance in the control of cell movements. For example, by anaylsis of time-lapse films of fibroblast behavior Abercrombie has shown that the movement of fibroblasts from the edge of an explant to form the "zone of outgrowth" is most likely due to contact inhibition. Cells at the periphery of the explant have some free surface at which a new ruffled membrane may form. This eventually provides sufficient momentum to pull the cell away from its fellows. As long as there is some cell-free space a fibroblast will move into it, and it will continue to move over the substratum until it either overtakes or is overtaken by another fibroblast. They will then "contact inhibit" each other but may break away again if a new ruffled membrane forms. At first, when cells leave the explant they will move outward. Soon, however, their direction of movement becomes random, unless there is continued emigration from the explant. With continuing emigration from the explant there are always cells behind and lateral to the first cells to have emerged. These orient the movement of the cells at the edge of the outgrowth into a substantially radial direction. As more and more cells leave the edge of the explant, there will be increasing probability of motile cells contacting each other. When the only surface available is that of its neighbor cells, a fibroblast will come to a standstill; it will not crawl over other fibroblasts. It is for this reason that cells in the zone of outgrowth tend to become arranged in an ever widening monolayer. The significance of contact inhibition for the behavior of cells within an organism received a heavy boost when Abercrombie was able to show that invasive sarcoma cells are not contact inhibited by fibroblasts in vitro. A front of fibroblasts constitutes no barrier for them; sarcoma cells crawl over fibroblasts as readily as over glass.

The means by which one fibroblast inhibits the activity of another is not well understood. The effect could be physical, with shearing forces between the two cells paralyzing the surfaces in some way. Or it could be primarily chemical. Taylor has noted that cells in culture are highly sensitive to pH changes, responding to increments as small as 0.2 pH units. A pH below 7.3 retards membrane activity, and when pH 5.8 is reached a cell is immobilized. Thus, contact inhibition could result from a highly localized accumulation of acid metabolites between two opposed cell surfaces. This interpretation could apply to contact inhibition between various kinds of fibroblasts, where there appears to be no specificity, but it is difficult to see how it would account for the lack of contact inhibition between sarcoma cells and fibroblasts. Because of their high rate of glycolysis, cancer cells tend to acidify their microenvironment.

Among the several possible mechanisms that have been proposed by Abercrombie, one has recently received substantial support. Carter has shown that the degree to which cells inhibit each other's movement on contact can be controlled by varying the substratum. Cells in culture do not readily adhere to a substratum of cellulose acetate, but if this surface is metallized in vacuo with palladium, cells will adhere to it and spread. By varying the amount of palladium deposited by shadowing, one can produce a range of surfaces to which cells presumably adhere to different degrees. In such a system the tendency of cells to form monolayers and show contact inhibition varies directly with the amount of metal on the substratum—the more palladium, the greater the tendency to adhere to the substratum and form a monolayer, hence the greater the contact inhibition; the less palladium, the greater the tendency of the cells to adhere to each other and form clumps, hence the less the contact inhibition.

These results are most provocative, for they suggest that contact inhibition is entirely dependent on the relative strength of cell adhesion to the substratum and not on a special reaction between the opposed cell surfaces. Thus, cells showing contact inhibition on glass would do so simply because they adhere more strongly to the glass than to each other. This hypothesis accounts very well for one of the features of contact inhibition—the tendency of cells to spread on the substratum rather than over each other. But it does not account for three other features—the paralysis of the ruffled membrane, contraction of the cell, and the formation of a new ruffled membrane—that occur when the surface of another cell is contacted.

Abercrombie has recently investigated this matter further by presenting cells with an agar substratum. Cells adhere to agar less than to glass and show a high degree of overlapping. When cells move up to the edge of the agar they stop. But the ruffled membrane continues its activity, suggesting that more than a change in the substratum is required for paralysis of the ruffled membrane. If we combine these results it seems possible that the most obvious features of contact inhibition may depend on different causes. Monolayering could be caused by the cells being more adhesive to the substratum than to each other and paralysis of the ruffled membrane and cytoplasmic contraction caused by contact with another cell surface. Further investigations now in progress will no doubt clear up this rather puzzling situation.

Adhesion to the substratum

It is obvious that the crawling movement of a fibroblast depends on adhesion to the substratum, since such cells move by a process of making and breaking adhesions, accompanied by contractions and relaxations. The adhesion needs to be firm enough for a cell to get a grip on the substratum and pull itself over it. But the adhesion must not be too strong, else cells could become completely immobilized. In a similar vein, too low a degree of adhesion would cause cells to slip, become rounded, and cease movement. It is clear therefore that a knowledge of the adhesive characteristics of cells is essential for an understanding of their locomotion.

The analysis of the relation between cellular adhesiveness and locomotion is still in its infancy. It is known, however, that moving cells are generally flattened on the substratum. If it could be established that degree of flattening were a direct measure of adhesiveness, we would have a convenient means of measuring the relationship between adhesiveness and locomotion. Actually, flattening might be due to several factors: increased adhesiveness, increased elasticity of the cell surface, lowered viscosity of the cytoplasm, and decreased strength of the cell membrane. But there is some crude indication that at least in some cases flattening is due to increased adhesiveness. Cells that are flattened on the glass in a tissue culture are less readily detached than more rounded cells when they are subjected to turbulence in the medium. Moreover, rounded anaplastic cancer cells are less adhesive to a fibrin substratum than their flattened fibroblast counterparts. In observing the degree of flattening cells, however, it is important that the conditions be the same. Visual inspection shows that the area of an individual cell facing the glass is greater when the cell lies near the edge of the outgrowth than when it lies close to the explant deep within the zone of outgrowth. Clearly, the degree of flattening of cells which presumably have the same degree of adhesiveness diminishes in regions of high cell density. Under such conditions they adhere more to each other than to the glass substratum. Actually the instruments are available for a quantitative study of the relation between flattening and adhesiveness. Detachment from the substratum, hence degree of adhesiveness to it, could be measured in the centrifuge microscope or by controlled turbulence. The degree of flattening could be revealed with a high degree of precision by studying the degree to which cells displace the fringe lines in an interference microscope. Until such studies are made we must rest on the impressions derived indirectly. These are not without their value, however. They have a consistency which supports their validity, and they pose some of the problems awaiting attack.

Cells are known to vary greatly in their adhesiveness to the same substratum. During the mitotic cycle they may adhere closely to their neighbors during interphase but usually round up and detach during mitosis, and then, after completion of the process, the daughter cells readhere. Cells also vary in adhesiveness during embryonic differentiation. Some cells which have low adhesiveness give rise to daughters with increased adhesiveness. Dissociated blastula cells of *Fundulus* (Teleostei), for example, adhere only slightly to each other and to glass, but their immediate descendants taken from an early gastrula cohere strongly and flatten on the glass. Primary mesenchyme cells of an early gastrula of a sea urchin apparently undergo a reduction in adhesiveness that causes them to round up and escape from the vegetal plate into the blastocoel. Once free in the blastocoel, however, their adhesiveness seems to increase again, and they crawl over the inner surface of the blastocoel wall. (In this case, of course, the adhesion between cells may not be due so much to an increase in adhesiveness of the mesenchyme cells as to their contacting a more adhesive substratum.)

Other cells which are adhesive and cling closely to one another during early phases of their differentiation become less adhesive as they mature. The erythrocyte is a good example of this. Erythroblasts cohere tightly, but mature erythrocytes are completely separable. The transformations that turn a normal tissue cell into a migratory, metastatic cancer cell appear to have changes in cellular adhesiveness as one of their essential features. The main evidence for this stems from an ingenious technique of pulling cells apart with glass needles. Application of this technique by Coman has revealed several strains of cancer cells to be less cohesive than normal tissue cells. It is of interest that these less adhesive cells tend to have rounded contours, abetting the conclusion that flattening on the substratum is associated with increased adhesiveness.

The adhesiveness of the cell surface is not the only feature in adhesion of cells to the substratum. The nature of the *substratum itself* is also important in determining whether a given cell will adhere to it and what the strength of the adhesion will be. In the older embryo and the adult much of the substratum for cell adhesion is collagen. Within the early embryo, cells usually serve as substrata for the adhesion of other cells; hence we would expect the nature of the substratum to vary with the different cell types and with their stage in development. This is indeed the case. Tissue cells of both early embryos (e.g., gastrulae) and organs in advanced stages of differentiation (e.g., kidney) show markedly different degrees of adhesion to different kinds of cellular substrata. Ectoderm of an amphibian gastrula adheres readily to mesoderm but not to endoderm. Similarly, endoderm adheres readily to mesoderm but not to ectoderm (see p. 102). Epidermis of advanced larvae adheres to other epidermis and to normally contiguous tissues like buccal epithelium and cornea through the intermingling of cells at the juncture. With esophageal and other epithelia normally not contiguous with epidermis, however, the adhesion is different and involves spreading of one epithelium over another.

Since we know that cells will not move unless they adhere to the substratum and that the adhesiveness between cells and their substrata varies, it would be of interest to know just how these variations affect the motility of cells. Aside from the possibility that an increase in adhesiveness in *Fundulus* (see p. 199) from blastula to gastrula is associated with acquisition of the capacity to move, we have had little to say on this matter until recently. Carter has now shown that tissue-culture cells (strain L) do not readily adhere to a cellulose-acetate surface but will adhere readily if this surface is metallized with palladium (see p. 25). By varying the amount of metal deposited it is possible to produce a range of surfaces that presumably allow different degrees of cellular adhesion. If palladium is deposited in a gradient, it may be seen that the cells on the gradient flatten on the substratum and move in the direction of increasing deposition of palladium. Cells on an evenly metallized surface move randomly.

Carter has also proposed that pseudopod or ruffled-membrane extensions are passive phenomena, which result only from spreading forces between the cell and the substratum. This is obviously not so in several cases where locomotory extensions have been observed to extend directly into the medium (filopodia of primary and secondary mesenchyme cells of sea urchins, spikes and filopodia of vertebrate tissue culture cells, filopodia of dissociated sponge cells cultured in sea water, and lobopodia of deep cells of a teleost blastoderm in culture and in vivo).

Contact guidance

Historically, the most impressive evidence that the substratum plays an important role in the movement of cells was provided by studies on the orientation of cells and of cell movement. Harrison, who invented tissue culture, had already noted in 1914 that frog neuroblasts do not move out into the liquid medium but prefer instead the surface of the glass culture vessel or other solid substrata. He then went on to observe that if the substratum is oriented it may play an additional role in providing orientation for moving cells. This important discovery has been investigated systematically and at length by Weiss, who claimed that cells may even gain orientation from the lining up of colloidal micellae in a protein film. Weiss termed this behavior "contact guidance." Clearly the association of a cell with its substratum is an intimate and possibly highly involved affair. This work is of great interest, not only as a possible explanation for the oriented cell movements that take place within the organism, but also because it set the stage for much contemporary analysis of the nature of the cell-substratum relation. It is therefore appropriate at this juncture to consider it in some detail.

Fibroblasts from the embryonic chick heart move so readily in culture that they have served as material for many of the studies to be described. This must be kept in mind, since it does not necessarily follow that all tissue cells will behave similarly. Certainly there are important differences between fibroblasts and epithelial cells, for example. The heart tissue in these studies is explanted to a clot of adult chicken plasma and chick embryo extract. The explant always digests the clot a little and imposes stresses which may line up the fibrin fibers in particular orientations. When two explants are placed near each other in the same clot, lines of stress are most pronounced between the two explants. This has a profound effect on the behavior of migrating fibroblasts. They orient preferentially in the in-between region to form a densely populated cellular bridge connecting the two explants (Fig. 2-4). A similar lining up of fibroblasts can be induced, if prior to explantation the fibers of the clot are given orientation by stroking with a camel's-hair brush. The fibroblasts stretch and take up positions parallel to the orientations of the substratum. Even more regimented formations occur if the cells are presented with a glass substratum which has been scored with fine parallel grooves. Individual cells assume exceedingly elongate spindle shapes in their efforts to conform precisely to the grooves (Fig. 2-5).

But why would cells orient on glass or plastic? We have no reason to

Fig. 2-4 Graph, giving the maximum distances reached by cells of the same cultures migrating in plasma clots and along glass fibers. Abscissa: days after explantation (with number of averaged cases). Ordinate: maximum migration in micrometer units (1 unit = 172 μ). –o–o– Cells on glass fibers, – –•– –•– – cells inside plasma. (Weiss, 1945. J. Exptl. Zool. **100**:353.)

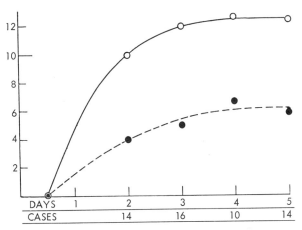

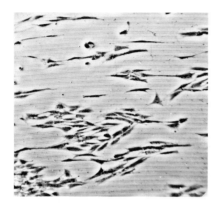

Fig. 2-5 Phase contrast picture of living culture of suspension of 8-day chick embryo skin cells on finely grooved glass. (Courtesy of Paul Weiss. Weiss. 1958. Intern. Rev. Cytol. 7:1.)

suspect that these materials offer orientations to which a cell could respond. Weiss suggests that macromolecular materials exude from cells and become oriented in the long axes of the grooves. The leading edge of a cell might then trace these tracks and orient the movement of the cell. It has been shown that cells do indeed release macromolecular materials which coat the substratum.

There is, however, an alternate explanation. Orientation in the groove could be simply the result of greater probability for contact of ruffled membranes, because more surface is available in the grooves than in the flat areas in between. If this turned out to be the proper explanation we could still say that cells are oriented by contact in the grooves but not necessarily because of micellar orientation of the contacted surface.

Some experiments of Rosenberg suggest yet another means by which cells may orient. He has shown that cells which adhere to one substratum, such as quartz, more readily than to another, such as a film of barium stearate-stearic acid, can detect the quartz, even though separated from it by several monolayers of the fatty acid. The technique was to coat a quartz slide with a predetermined number of monolayers of the fatty acid and cut troughs 10–100 μ wide in the film. After this, additional monolayers were superimposed to yield troughs 20–1000 A deep whose floors were 25 to several thousand angstroms distant from the quartz slide. Single-cell suspensions were then randomly dispersed on these hydrophobic surfaces. Cells tended to adhere preferentially in the troughs, even though the floors of the troughs were in all cases separated from the quartz surface by varying numbers of monolayers of stearic acid. This occurred even in shallow troughs only 60 A deep! In narrow troughs, spreading was polarized and individual cells demarcated the region by lengthwise alignment and orientation. In this instance, the cells were clearly guided by differential adhesion to the substratum and not by its orientations. As already discussed, Carter has shown that cells will tend also to adhere differentially to a substratum of shadowed metallic palladium and indeed will move up a gradient of increased palladium concentration. This likewise demonstrates oriented (indeed directional) movement in the absence of orientations of the substratum.

Weiss suggests that there may also be orientations of micellae of the colloidal ground substance in vivo that could provide ultrastructural guides for cell locomotion within the organism. Thus far, the intact organism has successfully resisted efforts to find indisputable evidence for such "contact guidance" by orientations of the ultrastructure, but many observations made by many investigators through the years find their readiest explanation in this hypothesis. Among them are the migration of pigment cells along blood vessels, the movement of nerve axons along blood vessels, myotome boundaries and the remains of degenerating nerve fibers, the posteriad movement of the lateral line organ and the Wolffian duct, the oriented protrusion of dendritic extensions of melanocytes along oriented rows of barbule cells in a feather germ, the movement of clusters of heart-forming cells on an oriented endodermal substratum, and the movement of neural crest cells ventrad along the side of the neural tube regardless of the orientation or location of the neural tube in the embryo. It appears therefore to be the best hypothesis yet available for giving orientation to the movements of most cells. Before leaving contact guidance, however, it must be emphasized that although it could give orientation to cell movements (north-south as opposed to east-west), it could not of itself provide directionality (north as opposed to south). Supplementary influences are required for this.

An understanding of the physicochemical nature of the adhesion of cells to each other and to intercellular materials is ultimately essential for an understanding of cellular locomotion and immobilization in the developing organism. At present, however, it is a highly speculative subject in which the results do not lend themselves to unequivocal interpretation. It is therefore best treated in a later section, when we deal with the difficult problem of the selective adhesions of cells.

Cell movements in vivo

We know little as yet about the mechanisms whereby cells move within the organism. Technical problems which impede effective use of phase and interference microscopy have been the main obstacles. All we can say at present is that all cells known to be migratory as individuals appear to be either amoeboid or fibroblastic in their type of movement. Amoeboid types with a conspicuous flow of cytoplasm include the leucocytes and lymphocytes of the circulatory system and the macrophages or tissue histiocytes. Cells which appear to move like fibroblasts are mesenchyme cells, neural crest cells, melanoblasts and melanocytes, and epithelial cells. Nerve fibers also appear to move out from the neuroblast cell body by a crawling movement of the extremity.

What impels a cell to begin movement within the organism, after a stationary existence, is unknown. This and its corollary—what causes such movements to cease—are two of the most important and fascinating problems. For example, what causes cells to begin moving away from the neural crest and then to cease movement when proper destinations are reached? What causes neurites to terminate in a highly specific way to make particular connections? What causes germ cells to lodge in the gonads? Cells in a stationary state can be artificially stimulated to begin movement in a variety of ways: a stimulus to the local formation of tissue; a stimulus toward malignancy; an injury, as in wounding; explantation to tissue culture; or dissociation and mixing of cells with cells of another type. Cells engaged in wound closure stop when they meet cells from the other sides of the wound. In tissue culture, contact-inhibiting cells stop when they contact the surface of other cells. Cells in mixed aggregates stop when they meet other cells of the same type. These artificial situations can give some ideas as to possible cues for starting and stopping within the organism and will receive considerable attention in later chapters.

The movement of individual mesenchyme cells in the transparent gastrula of the sea urchin has been studied in remarkable detail in recent years by Dan and Okazaki and by Gustafson and his co-workers with time-lapse cinemicrography. Just prior to invagination cells slip away from the vegetal plate and move into the blastocoel, where they soon arrange themselves in a ring around the indenting archenteron. These are the primary mesenchyme cells, which later give rise to the skeleton of the *pluteus* larva. Their liberation from their neighbors in the vegetal plate is accompanied by slow pulsating activity of their free inner surfaces, which results in the formation of rounded lobes. Most of the pulsatory lobes are confined to the free surfaces of cells bordering the blastocoel, but cells with several lobules can also be discerned. Pulsatory activity decreases when cells leave the blastocoel wall and pile up in the blastocoel. It would seem that a decrease in adhesiveness is the primary event. This would not only release cells from tight contact with their neighbors and the hyaline plasma layer (see p. 90), so that they could move into the blastocoel, but might also reduce tension in the cell membrane, so that the cells would now tend toward a spherical shape.

Free rounded surfaces of many cells tend to pulsate. For example, cells of many tissues from many organisms, when freed from their fellows by trypsin or EDTA, also become spherical and pulsate. In the case of the primary mesenchyme cells, the pulsations are associated with movement of the cells away from the vegetal plate into the blastocoel. When completely free, the rate of pulsation decreases, and the cells begin to send out long dendritic pseudopodia or filopodia. These pseudopods vary in length but may be as long as 30 μ. They protrude from the cell surface into the blastocoel. When they make adhesive contact with the inner surface of the blastocoel wall, they contract and pull the cell along, thus providing a basis for movement. Filopodia which do not make stable contacts or do not reach the blastocoel wall are eventually withdrawn.

The movement of these cells seems to be rather in the manner of an inchworm. It appears to be an exaggerated variant of the familiar crawling of fibroblasts, but here the protoplasmic contractions are clearly evident. The cablelike filopodia are never attached to the blastocoel wall except at their tips. Because of the transparency of the sea urchin egg this tip is easily accessible to detailed study and is found to have many branches, which sometimes appear to take their origins from an undulating sheetlike structure. Like the contacts of a fibroblast with the glass substratum, the contacts of these cells with the ectoderm are not permanent but continue slowly to break and reform, thus permitting changes in cell position. It is possible that local differences in adhesiveness of the blastocoel wall cause the mesenchyme cells ultimately to arrange themselves in their characteristic ring around the archenteron (Fig. 10-1). Some evidence in support of this hypothesis comes from the location of filopodial contacts. The exploring tips of the filopodia make contact selectively at the inner junctions between adjacent ectoderm cells, precisely at one of the points where these cells appear to adhere most firmly to each other (Fig. 2-6).

Fig. 2-6 Filopodial attachments of primary mesenchyme cells of *Psammechinus miliaris* to the ectoderm wall. The wall has been stretched and the pseupodial contacts at the cell junctions are clear. (Wolpert and Gustafson. 1967. Endeavour. **26**:85.)

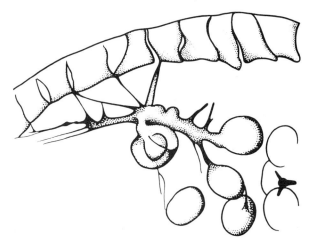

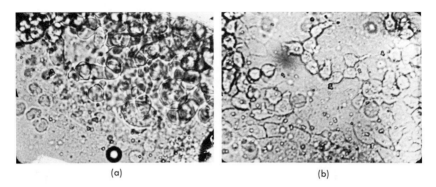

(a)   (b)

Fig. 2-7 (a) Deep blastomeres of a blastula of the teleost *Fundulus*. Most cells possess short lobopodia, two being apparent in profile. Translocation of the deep cells has not yet begun. The margin of the blastoderm where it joins the periblast stretches diagonally across the lower portion of the field. Three adjacent periblast nuclei may be seen at the lower left. Printed from time-lapse films of normally developing *Fundulus* blastoderms. Bright-field optics, ×100. (b) Deep blastomeres of a midgastrula (stage 12½) of *Fundulus* near the ventral blastoderm margin. An elongate lobopodium is adhering to another cell at the tip. Another blastomere possesses an adhering fanlike process, derived from a short lobopodium. These cells are actively motile. At the upper left, the outlines of the flattened cells of the enveloping layer may be discerned. (After Trinkaus and Lentz. 1967. J. Cell Biol. **32**:139.)

In a recent time-lapse study of cell movements during epiboly of the *Fundulus* blastoderm we observed a similar mode of locomotion by the deep mesentoblast cells. Prior to epiboly the cells protrude lobopodia that do not adhere and are withdrawn (Fig. 2-7a). Coincident with the onset of epiboly, however, the distal ends of the lobopodia begin to adhere to other cells. With this, some lobopodia are converted into stretched filopodia (Fig. 2-7b). These filopodia are contractile and usually shorten; the parent cell then moves in the direction of the shortening. Other lobopodia are converted into flattened fans when they become adhesive. These appear to function like the ruffled membranes of fibroblasts, leading the cell as it moves about. By these means deep cells translocate extensively and continually throughout epiboly.

SELECTED REFERENCES

Abercrombie, M. 1967. Contact inhibition: the phenomenon and its biological implications. Nat. Cancer Inst. Monograph. **26**:249–277.

Abercrombie, M. and E. J. Ambrose. 1962. The surface properties of cancer cells. Cancer Res. **22**:525–548.

Abercrombie, M. and J. E. M. Heaysman. 1953. Observations on the social behavior of cells in tissue culture. I. Speed of movement of chick heart fibroblasts in relation to their mutual contacts. Exptl. Cell Res. **5**:111-131.

Abercrombie, M. and J. E. M. Heaysman. 1954. Observations on the social behavior of cells in tissue culture. II. Monolayering of fibroblasts. Exptl. Cell Res. **6**:293-306.

Abercrombie, M., J. E. M. Heaysman, and H. M. Karthauser. 1957. Social behavior of cells in tissue culture. III. Mutual influence of sarcoma cells and fibroblasts. Exptl. Cell Res. **13**:276-291.

Allen, R. D. 1961. A new theory of amoeboid movement and protoplasmic streaming. Exptl. Cell Res. Suppl. **8**:17-31.

Ambrose, E. J. 1961. The movements of fibrocytes. Exptl. Cell Res. Suppl. **8**:54-73.

Carter, S. B. 1965. Principles of cell motility: the direction of cell movement and cancer invasion. Nature. **208**:1183-1187.

Goldacre, R. J. 1961. The role of the cell membrane in the locomotion of amoebae, and the source of the motive force and its control by feedback. Exptl. Cell Res. Suppl. **8**:1-16.

Gustafson, T. and L. Wolpert. 1961. Studies on the cellular basis of morphogenesis in the sea urchin embryo. Directed movements of primary mesenchyme cells in normal and vegetalized larvae. Exptl. Cell Res. **24**:64-79.

Harrison, R. G. 1914. The reaction of embryonic cells to solid structure. J. Exptl. Zool. **17**:521-544.

Ingram, V. M. 1969. A side view of moving fibroblasts. Nature. **222**:641-644.

Rosenberg, M. D. 1963. Cell guidance by alterations in monomolecular films. Science. **139**:411-412.

Weiss, P. 1945. Experiments on cell and axon orientation in vitro: the role of colloidal exudates in tissue organization. J. Exptl. Zool. **100**:353-386.

Weiss, P. 1961. Guiding principles in cell locomotion and cell aggregation. Exptl. Cell Res. Suppl. **8**:260-281.

Wolpert, L., C. M. Thompson, and C. H. O'Neill. 1964. Studies on the isolated membrane and cytoplasm of *Amoeba proteus* in relation to ameboid movement, p. 143-171. *In* R. D. Allen and N. Kamiya [ed.] Primitive motile systems in cell biology. Academic Press, New York.

THREE

Directional Movements and Chemotaxis

During development there are many instances of cells moving directionally as individuals and accumulating in specified regions. Cells of the germ ring of a teleost gastrula converge dorsally and eventually form the embryonic shield. Primary mesenchyme cells of a sea urchin gastrula migrate along the inner surface of the blastocoel wall to form a ring about the invaginating archenteron. Neural crest cells migrate from the dorsal surface of the neural tube and move to their several destinations in prescribed directions along defined routes. Amoebae of the cellular slime molds form streams which converge on centers, to which they finally all migrate. If a nerve is cut, the fibers that regenerate from the stump move out along the path defined by the degenerating cut fibers. In the egg of the Japanese medusa, *Spirodon*, the fertilizing sperm always enters near the nucleus, and excess sperm are always found near the egg nucleus. Pigment cells of certain newts space themselves equidistant from each other in the flank skin. Leukocytes accumulate at a focus of inflammation. Sperm of a bracken fern move toward and accumulate in the archegonial pores. Clusters of precardiac cells migrate anteromedially in the chick blastoderm to converge and form the heart primordia. We could multiply the examples, but these are sufficient to demonstrate that the movement of cells in defined directions to accumulate at particular loci is a widespread phenomenon of great developmental and functional significance.

Trap action

We have already discussed the possible importance of the substratum in giving orientation to cell movements. Contact guidance is a likely means by which moving cells may gain orientation in the organism. But since it cannot give direction it is not a sufficient mechanism for guiding cells to a particular locus. A gradient of increased adhesiveness of the substratum, however, could give direction to motile cells and seems a possible normal means of achieving it. Unfortunately, the only concrete evidence in its support comes from a totally in vitro system—that of Carter (p. 28), who showed that L-cells will move up a gradient of increasing palladium deposition on the substratum.

Cells might also accumulate in significant numbers at a given place after completely random movements, if they are selectively eliminated elsewhere or if there is some means of trapping those that by chance enter this region. One can imagine several possible mechanisms for such *trap action*: selective adhesion, slowing down or cessation of movement, and presence of a more favorable nutritional environment that stimulates cell multiplication. Selective adhesion and other mechanisms of trap action will be discussed at length later, in connection with the mechanism whereby cells sort out from each other in mixed aggregates and arrange themselves during gastrulation, so we will not dwell further on them here. It should be pointed out, however, that it is often a difficult matter in practice to distinguish between trap action and chemical attraction over short distances. Where exceedingly short distances are involved the distinction may be semantic.

Chemotaxis

This brings us to a consideration of *chemotaxis* as a means whereby cells may both acquire directional movement and accumulate at a common destination. Chemotaxis is involved when the direction of movement of cells is influenced by a gradient in concentration of substance(s) in solution. In spite of widespread conviction to the contrary, it is not an easy matter to determine whether chemotaxis is at work. It is not sufficient, for example, to observe only the end result, that cells accumulate at a given locus. They may be trapped there, be guided there by an oriented substratum, or simply be wafted there passively by convection. It must be shown that the *direction* of movement of individual cells has actually been influenced. Moreover, the movement must be concentration dependent. A cell must be able to detect a concentration difference over a distance corresponding to its own diameter and move parallel to the gradient. In addition, contact guidance or other influences of the substratum must be excluded.

A classical method of testing for chemotaxis, for example, involves inserting a capillary tube containing the suspected chemotactic agent into a medium in which responsive cells are present. If cells accumulate in the tube, it is then often thought that chemotaxis is at work. But cells may be drawn to the tube by convection currents resulting from physicochemical differences between the medium inside and that outside the tube or by a pumping or suction action due to capillarity. Or, cells moving into the tube by chance may not readily leave because of one or another kind of trap action, such as the mouth of the tube being too small to permit random movement. Because of insufficient attention to pitfalls like these there are not many claims of chemotaxis which exclude other interpretations. As a consequence, I will ignore the great bulk of the work on chemotaxis. There are in fact so few unequivocal examples of chemotaxis that there is space to discuss almost all of them.

Bracken fern spermatozoa

One of the best known was first studied by Pfeffer in 1884. A random suspension of bracken fern spermatozoa (antherozoids) accumulates at the archegonium. A long series of experiments finally led Pfeffer to conclude that chemotaxis is at work and that malic acid is the attracting compound. He filled a small glass capillary with sodium malate and inserted it into a suspension of sperm. Sperm not only accumulated in the capillary but showed a precise orientation toward the mouth of the capillary in the concentration gradient. There had to be a ratio of 30 between the original concentration of sodium malate in the capillary and in the sperm suspension. Since the direction of sperm movement was clearly influenced in this experiment there is little question that chemotaxis is operating. However, the experiment does not exclude the possibility that other factors cooperate to increase the accumulation of sperm in the capillary.

It is an interesting commentary on the uneven progress of science that it was not until 70 years later that the critical experiments were done. Rothschild filled a glass pipette with 1% agar containing a 1% solution of sodium malate. The solid agar eliminated the possibility of capillarity and hydrodynamic flow. He then made cinematographic records of sperm movement. Under these ideal experimental conditions individual sperm still oriented toward the open end of the capillary and moved to it (Fig. 3-1). Another study by Brokaw has shown that the attraction is not a statistical affair but that the direction of movement of *individual* spermatozoids is influenced by the presence of the compound. He also found that

Directional Movements and Chemotaxis 39

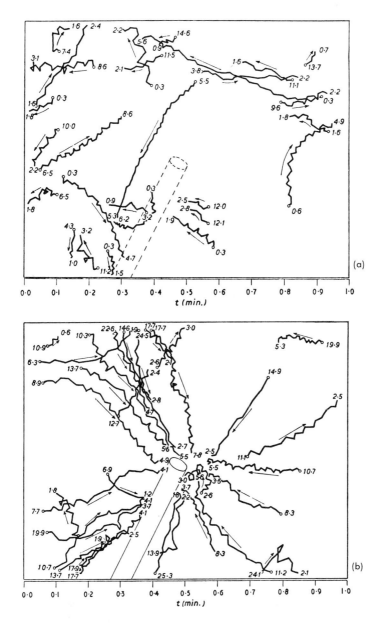

Fig. 3-1 (a) Movements of bracken spermatozoa in tap water. The circles indicate where each track begins. Numbers at the beginning and end of each track refer respectively to time of start and duration of tracks in seconds. (b) Movements of bracken spermatozoa after insertion of pipette (diameter 30 μ) containing 1% sodium L-malate in a 1% agar-tap water gel, into the same sperm suspension as in (a). Numbers at the beginning of tracks indicate time in seconds after insertion of pipette. (Rothschild. 1956. Fertilization. John Wiley & Sons, N.Y. 170 p.)

malic, maleic, and citraconic acids produced the chemotactic effect, whereas related compounds such as fumaric, mesaconic, dihydroxymaleic, or malonic acids did not. It appears that the bimalate form of malic acid is chemotactically most active. Thus the old conclusion is justified. Bracken fern spermatozoa are chemotactically attracted by malic acid. It must be emphasized, however, that this is not the only possible chemotactic agent, nor is it yet known what the normal agent is in the archegonium.

Animal cells

There have been many claims of chemotactic behavior by animal cells, based largely on the end result—an accumulation of cells. One of the first claims was by none other than Wilhelm Roux, whose bold experiments on amphibian eggs in the last years of the 19th century helped found the science of analytical embryology. He teased apart the cells of early amphibian embryos and observed that they tended to aggregate in culture. He assumed this to be due to "cytotropic" factors released by the cells and concluded that such factors could play an important role in morphogenesis. This conclusion remained unchallenged for decades, no doubt in part because of the eminence of its originator. It was finally reexamined in the 1930s, fittingly by two other German embryologists, and found to be groundless. By painstakingly following individual cells in culture, they found their movements to be entirely random. Such has been the fate of all other suspected cases of chemotaxis of animal cells which have been subjected to rigorous scrutiny, with four exceptions: the mutual repulsion of amphibian propigment cells, the mutual repulsion of leukocytes, the movement of polymorphonuclear leukocytes toward certain test objects like bacteria, and the attraction of sperm to the female gonangium of hydroids.

The idea that melanoblasts or propigment cells of newts might mutually repulse each other by a form of negative chemotaxis first came from the observation that during the primary phase of pigment pattern formation the pigment cells are distributed more or less uniformly on the flanks of the larvae. The first experiment performed to test this hypothesis consisted in transplanting neural crest (from which pigment cells originate) to an area of the flank which had been made pigment free. Propigment cells emigrate radially from such a graft in all directions, but they do not invade territory already occupied by pigment cells. This result led eventually to a direct test in cell culture, in which the only cells present were pigment cells. The cells were confined in fine capillary tubes and their motile behavior observed. If their distribution in vivo is due to diffusible substances produced by the pigment cells themselves, two or three propigment cells should move apart in the capillary, and one isolated cell should show little movement. Twitty and Niu, who performed these experiments, found that two or three cells do in fact mutually repulse each other in a strongly directional way, moving apart a certain distance and then ceasing movement. Single isolated cells, in contrast, move much less, never wandering over a distance greater than a few cell diameters (Fig. 3-2). These beautiful

Fig. 3-2   Newt pigment cells isolated singly (a) in capillary tubes remain essentially stationary, whereas those isolated in pairs or larger groups (b and c) move away from one another in response to diffusible substances released by the cells. Arrows indicate individual flattened cells. (Courtesy of Man Chiang Niu. After Twitty and Niu. 1954. J. Exptl. Zool. **125**:541.)

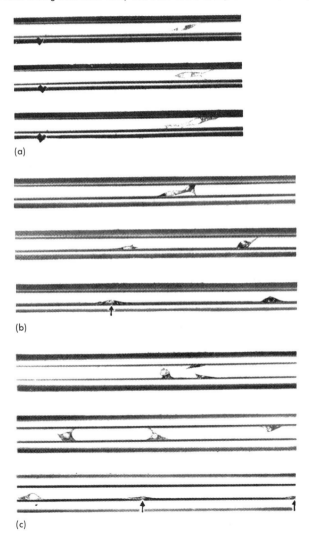

experiments lend strong support to the hypothesis that the initial phase of pattern formation during normal development is due to the negatively chemotactic activities of the pigment cells themselves. The active substance(s) has not been identified. During later phases of pattern formation pigment cells of one species pull together again, and their arrangement appears to be influenced by other cells.

It is also possible that white blood cells may be negatively chemotactic toward each other. The region between two colonies of polymorphonuclear leukocytes in culture is usually devoid of cells. This so-called "no-man's-land" effect is most readily explained as due to negative chemotaxis. If chemotaxis is indeed at work, however, it must be rather weak, for it is ineffective in inducing repulsion between individual cells. To be detectable it requires the influence of a whole colony.

That leukocytes and lymphocytes are chemotactically attracted to a focus of inflammation has of course been suspected for a long time. It is only in recent years, however, that each of these types has been examined under controlled conditions. The method was to culture the cells under a coverslip in a defined medium in which the viscosity had been increased by a biologically inactive polysaccharide, polyvinyl pyrrolidone, so that no detectable convection occurred. The test object was usually a clump of bacteria such as *Staphylococcus albus*. Under these conditions, the cells adhere to the glass substratum and move in converging streams toward the test object. Since individual cells can be seen to move on the glass surface by active amoeboid movement toward the test object, the possibility of their being wafted there by movements in the medium is eliminated. Inasmuch as no fibrin strands were observed to form in a radial fashion about the test object, contact guidance seems a most unlikely possibility. By such experiments, polymorphonuclear leukocytes, eosinophil leukocytes, and monocytes have been shown to be chemotactically responsive. Lymphocytes, however, show no chemotaxis under these conditions. Incidentally, a wide variety of microorganisms and certain other complexities, such as starch grains, have chemotactic activity for all three kinds of responsive cells. These studies have given no leads as yet as to why certain lesions result in the accumulation of polymorphonuclear leukocytes and others in the accumulation of monocytes or eosinophils or lymphocytes. As far as mechanism is concerned, all we can say is that the agent(s) presumably have an effect on the sol-gel state of the cytoplasm, such that pseudopodal formation in a certain direction is favored. Until defined chemical agents are isolated, little further progress can be anticipated in the study of mechanism.

Experiments which appear to lay a basis for substantial progress in this direction have been performed recently in Boyden's laboratory in Canberra. The key to this research lay in the elaboration of an in vitro technique whereby the chemotactic movements of polymorphonuclear leukocytes could be assessed quantitatively. In essence, it consists of interposing a millipore filter with a pore size of 5 μ between two compartments, one containing leukocytes in normal serum and the other containing the soluble presumed chemotactic agent. The pore size is large enough to permit leukocytes to squeeze through. To assess the results after a given period of time, the number of cells is counted on the far side of the membrane. Some definitely encouraging progress has been made by the use of this method. In the first place, when an antigen, the bacterial toxin tuberculin, was present in the test compartment, the number of leukocytes found on the far side of the membrane was substantially higher than in the control, indicating a strong chemotactic effect. But a potentially even more interesting result was obtained when antigen-antibody mixtures were used. Incubation of such a mixture in fresh normal serum at 37° caused strong chemotactic attraction of leukocytes. Further analysis revealed that it is not the antigen-antibody complex itself which is the chemotactic agent, but a heat-stable (56°) product produced when the complex is incubated in fresh serum. Since the heat-stable agent is not produced in serum that has been heated at 56°, it appears that it is formed by the intermediary of a heat-labile system normally present in serum. Presumably, the antigen-antibody complex activates this heat-labile system, which in turn forms the heat-stable chemotactic agent. Since it is reasonable to suppose that antigen-antibody complexes can exert a chemotactic effect in vivo as well as in vitro, these results have obvious implications for the normal mechanism whereby polymorphonuclear leukocytes may be mobilized at a focus of bacterial infection within the organism. Inasmuch as the chemotactic product is heat stable, it is no doubt a small molecule and thus may soon be isolated and identified.

The fourth case of confirmed chemotactic behavior by animal cells was described just a few years ago. It involves the prefertilization behavior of the sperm of the hydroid *Campanularia*. In these organisms the eggs are encased in a gonangium whose single opening must be traversed by the sperm if fertilization is to occur. The behavior of sperm in the vicinity of a female gonangium was studied cinemicrographically, so that the movements of individual sperm could be tracked. Sperm swimming within a certain distance of the aperture of a gonangium were found to undergo changes in their mode of swimming which bring them closer to the aperture and eventually cause them to enter it. These changes are 1) radical alterations of direction with the result that subsequent movement is toward the aperture (Fig. 3-3) and 2) a definite increase in speed as they move closer to the aperture. Sperm of two different species of *Campanularia* were found to be chemotactically sensitive but mainly to their own female gonangia.

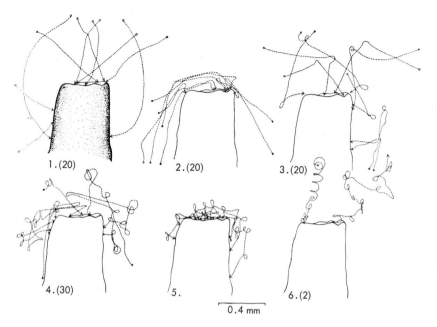

Fig. 3-3 Plotted trails of *Campanularia* sperm attracted to the distal end of a mature female gonangium. The circles represent the start of the trail. The figures in parentheses represent approximate percentages of the types of trails where the sperm approach the gonangium from a distance. Those trails which show similar characteristics have been placed into five rather rough categories. The first category consists of relatively straight paths, most of which would normally have led to entry. The second is the reverse of the first, in that the movements of the sperm should have taken them directly away from the aperture. However, as each sperm approaches, it turns toward the opening, moves across the aperture parallel to the distal end of the gonangium, and eventually turns abruptly into the opening. The third category consists of those sperm whose direction of approach is such that they would pass by the gonangium at some distance, moving roughly perpendicular to its long axis rather than parallel to it. In this case, sharp turns are made toward the gonangium even at some distance from the aperture (about 0.4 mm in the figure). The result is either entry through the aperture or at least collision with the distal portion of the gonangium. The fourth type is more complicated and shows portions of the three types already mentioned as components of each trail. Although many turns are made, apparently the result of indiscriminate spinning, the final direction taken up is consistently toward the gonangium and eventually leads to entry. The fifth case consists of those sperm already within the gonangium that attempt to swim out. In every case but those shown as category six, these attempts fail, and the sperm either turn around and re-enter the gonangium or else turn out over the lip of the aperture and strike the sides of the gonangium. The two examples in part six where the sperm escape represent about 2% of all the trails analyzed. The highly erratic nature of these paths, even when compared with those of category four, is obvious. (Miller. 1963. J. Exptl. Zool. **162**:23.)

The chemotaxis is thus species specific in contrast to ferns, where interspecies discrimination does not exist. Among the extracts of various tissues tested for activity only a tissue of ectodermal origin associated with the aperture of the gonangium yielded material which both activates and attracts sperm. The active principle is heat stable, nonvolatile, dialyzable, polar, and appears to be a single molecule of less than 5000 mol. wt. When capillary tubes containing agar impregnated with this substance are placed into sperm suspensions the sperm are seen to move toward the tube opening with considerable accuracy, the accuracy increasing with increased concentration of the attractant. This is the first rigorous demonstration of chemotaxis during fertilization of an animal. Since, however, there are other species in which swimming sperm must penetrate a small orifice in order to fertilize eggs, it will not be surprising if other similar examples of chemotaxis are soon discovered.

The cellular slime molds

Far and away the best analyzed case of chemotaxis is that of the cellular slime molds. We lack sufficient space to do justice to the many outstanding studies and so must limit ourselves to what is really a brief summary. At any rate the literature is critically summarized in several excellent reviews. They can be consulted to add substance and contour to the skeleton of this presentation.

The so-called cellular slime molds, or Acrasiae, have a remarkable life cycle. The cells move back and forth between a nonsocial phase, in which cells are separate and solitary, and a social phase, in which they are aggregated to form a multicellular organism. During the first phase the cells move about as individuals and feed on bacteria. Then they aggregate into clumps as they enter the second phase. This leads later to the differentiation of different cell types and the erection of fruiting bodies.

The process of aggregation is truly spectacular and has attracted the attention of biologists for years. The orientation of cells toward a center and the formation of streams of actively moving cells which converge on the center until finally all have moved together to form an aggregate had been thought for decades to be caused by chemotaxis. It remained unproved, however, until the studies of Bonner. In 1947 he first established the chemotactic nature of the aggregation by showing in a series of elegant experiments that amoebae of *Dictyostelium discoideum* are attracted to a center by a diffusible agent produced and emitted by cells in the center. First, he eliminated the possibility that attracting rays of some sort (such as "mitogenetic rays") are emitted by the center by placing a center on one side of a thin glass shelf suspended in water and separate amoebae on the underside. After a time, the amoebae all turned the edge of the shelf to the upper side and streamed into the center (Fig. 3-4). If rays had been involved and had penetrated the glass, the amoebae would have been gathered beneath the center on the other side of the glass. If the rays had not penetrated the glass, the amoebae would not have been affected by the center. No ray can be expected to round a corner.

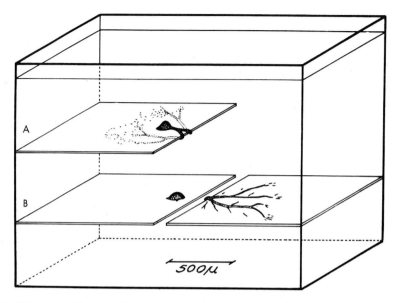

Fig. 3-4 A semidiagrammatic representation of two experiments done on aggregation in the cellular slime mold *Dictyostelium*, using coverslip shelves held under water. A, the myxamoebae previously at random under the coverslip are attracted around the edge to the center on the upper surface; B, the myxamoebae previously at random on the right hand coverslip are attracted to the center on the left hand coverslip, across the substratum gap. (Bonner. 1947. J. Exptl. Zool. **106**:1.)

Another possibility is that the center might exude an extracellular material which would coat the substratum around it and conduct the amoebae to the center by a form of contact guidance. This hypothesis was tested by placing two glass shelves side by side with a narrow gap between them (Fig. 3-4). A center was placed on the top of one and separate amoebae on the other. In spite of the gap, the separate amoebae streamed to the edge of their shelf to a point nearest the center and accumulated there. This proves that the attraction can cross a region lacking a substratum. The amoebae are attracted but cannot bridge the gap because they require a solid substratum for their locomotion. Actually, the amoebae can be seen "pawing the air with their pseudopods" in an apparent effort to cross the gap. But they do not succeed unless the shelves are pushed very close together. Then, with the gap only an amoeba length, the cells will form a hanging bridge across the gap to reach the center. By this means contact guidance was eliminated.

Bonner next showed that positive evidence for a diffusible chemotactic substance can be readily obtained. By gently flowing water over aggregating amoebae, he demonstrated that only the amoebae downstream oriented toward the center. Upstream amoebae showed no interest. He called the chemotactic substance(s) *acrasin*. This work of Bonner is the classic in the field, and it led immediately to intensive study of the chemotactic process in *Dictyostelium* and other slime molds in a number of laboratories. The leaders have been Bonner, Gerisch, Raper, Shaffer, and Sussman (in alphabetical order). Their papers and those of their students and associates should be examined by those who wish details of evidence and controversy; there is much of both.

It was thought for a long time that myxamoebae in the solitary phase are wholly independent and either move at random or show no evidence of directed movement. However, when Samuel took the trouble to carefully track the movements of cells during this phase, he found that their movements are in fact strongly influenced by bacteria and by each other. Movement of amoebae toward bacterial clumps is strongly positive. Cells may move toward and collect over a bacterial colony, even though separated from it by a layer of agar. In the absence of bacteria, myxamoebae distribute themselves evenly over the substratum and remain entirely separate from one another, unless the cell density is so high that contact cannot be avoided. Plotting successive positions of dispersing cells shows them to have a strongly negative chemotactic index. The cells appear to actively repulse one another like the pigment cells of *Triturus*. Myxamoebae will move only over wetted surfaces, but there is no evidence that like fibroblasts they follow scratches in glass or glass fibers. In spite of their name, solitary amoebae of the cellular slime molds apparently do not exude slime onto the substratum.

As long as there is a supply of bacteria, the amoebae feed and divide and remain in the solitary phase. But when the supply of accessible and edible bacteria is depleted, the antisocial responses of the cells are gradually replaced by social ones. In the species *Polysphondylium violaceum* the first evidence of this is the orientation of cells toward one cell in their immediate vicinity. Shaffer observed cells to orient toward a single, round, stationary cell (Fig. 3-5), which he called a "founder cell." This cell may be of any size. Once a founder cell acquires the property of attracting other cells it apparently retains it for some time, even a few hours. If a young center is dispersed, cells may reaggregate radially around the original founder cell a hundred times or more. If a founder cell is killed, the other cells will not reaggregate until and unless a new one arises from amongst them. It is not known what causes a particular cell to become a founder cell, but since there is no evidence that they preexist as such in the population it is assumed that any cell can become a founder cell, when exposed to the proper conditions.

This is in contrast to the ideas of Sussman, who, with his associates, has amassed an impressive body of evidence in support of a stable, latently

48 *Directional Movements and Chemotaxis*

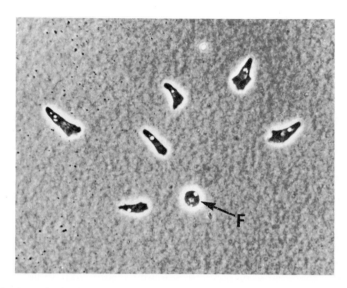

Fig. 3-5 A round *Polysphondylium violaceum* founder cell (F) attracting all its neighbors. (Courtesy of Brian Shaffer. Shaffer. 1962. Advan. Morphogenesis. **2**:109.)

different "initiator cell," which preexists in populations of *Dictyostelium discoideum*. These "I-cells" also differ from founder cells in that they are alleged to be larger than other cells and to constitute a fixed proportion of the population. The number of founder cells is greatly dependent on population density. The arguments for and against I-cells are long and complicated, and their existence is the subject of much controversy. Sussman claims, on the basis of a number of acute observations, that they are indispensible for aggregation; but other workers find that aggregation may occur in the absence of cells that answer to their description. For these reasons we will not discuss them further. The presence and role of founder cells in *P. violaceum* seems established, but it is not yet known whether cells like them initiate aggregation of other species. In three species of *Dictyostelium*, cells form small clumps before they show sustained radial orientation. Perhaps in these species the first body to be attractive, and therefore to form a center, is a small cluster of adhesive cells.

Whatever may be the precise mechanism of initiating a center, it is the emission of a chemotactic factor that causes solitary cells to move toward it. Actually, amoebae are already sensitive to the attractor before a natural source of it develops. They show this sensitivity first by a preferential cytoplasmic outflow on the side of the cell or the side of a pseudopod facing the source. Bonner showed that if the acrasin gradient is reversed, elongate cells will usually make a U-turn. Sometimes they round up and put out a new pseudopod. Or, indeed, a cell may extrude a new pseudopod from the rear and reverse its direction of movement, while still remaining elongate. Once cells have changed their direction of movement and move toward the center, it would be logical to expect that their speed would increase. There is surprisingly little evidence on the matter, but apparently there is such an acceleration in a case recently studied by Samuel.

In *Dictyostelium* a center can attract cells only for a distance of about 100 μ. Yet a stream of cells radiating from a center may extend up to 10 mm! This suggests that cells in the streams must also produce acrasin. This would explain an old observation that cells on one side of a dialysis membrane duplicate not only aggregation centers on the other side but the cell streams as well. Shaffer has in fact shown that there is no detectable decrement in acrasin concentration centrifugally along a stream. Apparently, when cells fall under the influence of acrasin, they too begin to produce it. Individual amoebae attracted toward a stream from its flank arrive at right angles and only gradually turn toward the center. This presents a problem. If acrasin will attract cells to the stream and acrasin concentration is approximately equal along the stream, how then do we explain the centripetal movement of cells in the stream? Shaffer notes that cells orient and move in the same direction as the *moving* cells to which they adhere. Cells streaming toward aggregation centers tend to stick together end to end in chains. The tip of a cell that approaches such a chain from the side receives no guidance from it, if it contacts the middle region of a cell in the chain. However, when the back end of the contacted cell moves past it, it turns and moves in the direction in which the chain is advancing. Shaffer calls this "contact following" and suggests it as the normal manner of adding cells to the stream. If the stream is not moving when newcomers adhere, they may turn either toward or away from the center. They invariably move toward the center only when the cells to which they adhere are themselves moving toward the center. Incidentally, contact following could be explained if an amoeba's surface remains stationary along its sides relative to the substratum and is created at the front end and removed at the back end. It is clear that whatever is going on in contact following, it is not the same as contact guidance. The latter makes use primarily of nonmoving substrata and is not polarized. Contact inhibition of course is not at all at work in these moving streams. What we have instead is a form of contact promotion, or rather promotion of locomotion by contact.

Once a center is established, it attracts cells only over relatively short distances. Yet, as aggregation proceeds it yields elaborate and beautiful radiate patterns, which extend far beyond the effective range of the acrasin produced by the center (Fig. 3-6a, b). In so-called stippled aggregation, characteristic of *Dictyostelium discoideum,* the form most studied, the cells

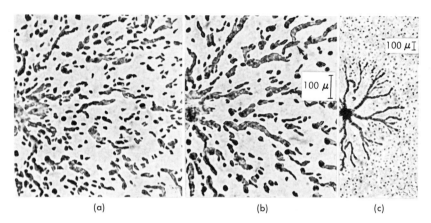

Fig. 3-6 (a) A stippled aggregation in *Dictyostelium*, the center being on the left. (b) The same field as Fig. 3-6a about ½ hr later. Arrowed: a short length of stream that has been attracted by a length ahead and to one side of it and has made a right angle bend to join it while the latter is in the process of responding in exactly the same way. Close inspection will reveal similar behavior throughout the area. (c) *Polysphondylium violaceum*. Continuous streams (the breaks are secondary) growing outwards from a center. Cells just beyond their outer ends are completely separate. (Courtesy of Brian Shaffer. After Shaffer. 1962. Advan. Morphogenesis. 2:109.)

become oriented when still largely separate from one another. They then gradually join each other to form short chains; these in turn merge into continuous streams, which by and large move directly toward the center. The fact that not all streams point directly toward the center is significant. Neither do all separate cells. Cells lie on arcs of varying curvature that spread out from the center. In extreme cases these arcs may curl around so strongly that some cells quite close to the center actually move away from it. How then explain the overall attraction of distant cells to the center? Why indeed is there a center? As a matter of fact a rather large amount of aggregation could occur in the absence of a center. If cells continued to move and to adhere to each other, they would form clusters, and many of these would eventually fuse to form large aggregates. Behavior like this is observed in the sorting out of vertebrate tissue cells from each other in mixed aggregates (Ch. 7).

We have already pointed out that no gradient in concentration of acrasin from the center is detectable. Acrasin is produced in apparently equal amounts along the length of each stream. Actually, it is easy to see, if such a gradient did exist, how it would present such small differentials in concentration that the responses of cells to it could be easily upset by many factors. By and large, however, once started, aggregation moves steadily to completion. A clue to what appears to be an adequate explanation came from Bonner's time-lapse films of aggregation. These show aggregation to occur in waves of abrupt orientation and rapid centripetal movement. These waves move outward from the center at a rate of about 0.5 mm a minute. This and other features of aggregation may be explained if we assume that, instead of extending over the entire field, the gradient is compressed into a narrow zone moving across it. This would expose cells to a differential in concentration large enough to be relatively independent of local variations.

Let us assume that a secreting cell stimulates its more centrifugal neighbors to secrete acrasin and become adhesive and that a cell attains maximal secretion rapidly and relatively independently of the concentration of the inducer. Thus the production of acrasin by the induced cells need not be related to the distance from the center nor greater at the center than elsewhere. The advantage of such a moving front of secretory activity is that it could relay a center's influence indefinitely with great speed and economy of material. It has been shown that the same cells can in fact be successively reoriented by different centers and could therefore propagate secretion fronts repeatedly. But, in order for successive fronts to be propagated, it is necessary to make one additional assumption: there must be a refractory period, during which there is a decrease in responsiveness after each front has passed. This would prevent cells from responding to the inducer released by more peripheral cells, which they had just induced, and would therefore prevent the front from reversing its direction. This implies that secretion also declines about as rapidly as it increases, i.e., that the fronts are pulses. Since the aggregating cells not only induce each other to produce acrasin but also cause an increase in adhesivity, the cells form chains and streams of increasing length. Once in the streams, they would be oriented both by successive waves of acrasin produced by the cells centripetal to them and perhaps by their contact relations as well. But what causes the center to emit the pulses of acrasin that trigger each successive wave of acrasin production is not established. Presumably the stimulus comes from the first cells stimulated by the center. After the cells of the center have emerged from the refractile period they may respond to the lingering remnants of acrasin produced centrifugally to them and emit a new pulse, which in turn would propagate a new radiating wave of stimulation and acrasin production. Although several important details of this hypothesis await verification, there is sufficient evidence now at hand to permit its tentative acceptance.

It is not so easy to explain the chemotactic mechanism in a form such as *Polysphondylium violaceum*, which possesses continuous streams (Fig. 3-6). Shaffer suggests that in this species cells are not stimulated to become both strongly secreting and adhesive *until* they actually make contact with a stream or center. Time-lapse films reveal no pulses.

Perhaps the most important of the many unknowns which yet remain is the problem of how a secretion front is started in the first place. How does a center originate? Whether it be initiated by I-cells, by founder cells, or other mechanisms, what causes the center initially to produce enough acrasin to trigger the first front? There are many speculations, but either they are not sufficiently supported by evidence, or the evidence is contradictory. It is a problem of importance and of course of interest to all slime-mold workers. But the solution will not come easily. Where one is dealing with the secretions of single cells or tiny groups of cells or the induction of secretion in them, the technical challenge is extreme.

Finally, a word about the efforts to define acrasin chemically. Acrasin may be fractionated functionally into several components: an attractor, an inducer of the attractor and an inducer of the inducer, and another inducer of increased adhesiveness. It is, as Bonner defined it, a "type of substance consisting of one or numerous compounds responsible for stimulating and directing aggregation."

Acrasins have varying degrees of specificity. *Dictyostelium* acrasins appear to be very similar and attract cells of other species within the genus. *P. violaceum* amoebae, on the other hand, aggregate separately when mixed with *Dictyostelium*. This suggests that their acrasins are different. Specificity like this usually depends on large molecules. Yet the conformation of aggregating patterns, when separated by a dialyzing membrane, suggests that the chemotactic factor is composed of small molecules. This was confirmed by extractions. Only by removing large molecules from the preparation was an active preparation possessing stability obtained. Sussman and Shaffer both found this to be true. Then Sussman and his co-workers went on to prepare a stable attractor by acid extraction which could be fractionated into a component active by itself, an inhibitor, and an inhibitor of the inhibitor.

A disturbing note was sounded when Wright and Anderson discovered that pregnancy urine and various pure steroids in very low concentration were also active in attracting myxamoebae. Since steroids are known to be experimentally active in many situations in which they are not normally present, the discovery suggests that some aspects of aggregation can be stimulated by a number of compounds and makes one wonder just how specific the extracts from the aggregation centers are. Curiously, even though the above discoveries were made in the middle 1950's, little was reported since, until very recently. It was obviously a subtle business, tantalizing in the quick preliminary successes but confusing and contradictory as the investigations were pursued.

Oftentimes in science a situation which at first appears highly complex becomes reduced to a few simplicities, as it becomes disentangled by research. Chemotaxis in the cellular slime molds seems to have moved in the opposite direction. What appeared to be an elegantly simple story

15-20 years ago has with detailed and clever analysis become highly complex. It now looks as if different species operate differently. Furthermore, with the close attention that is now being given to behavior of individual amoebae, it is becoming more and more evident that myxamoebae behave very differently during different phases of the cycle and move not only in response to chemotactic agents but also in terms of their contact relations with other cells. It seems likely that the complete explanation will ultimately involve a relatively small number of discrete factors. But to find out what they are and how they interact in space and time will require even more attention in the future than these wondrous creatures have yet received.*

SELECTED REFERENCES

Bonner, J. T. 1947. Evidence for the formation of cell aggregates by chemotaxis in the development of the slime mold *Dictyostelium discoideum*. J. Exptl. Zool. **106**:1-26.

Bonner, J. T. 1959. The cellular slime molds. Princeton Univ. Press, Princeton, N.J. 150 p.

Boyden, S. 1962. The chemotactic effect of mixtures of antibody and antigen on polymorphonuclear leukocytes. J. Exptl. Med. **115**:453-466.

Brokaw, C. L. 1958. Chemotaxis of bracken spermatozoids. J. Exptl. Biol. **35**:192-196.

Gerisch, G. 1968. Cell aggregation and differentiation in *Dictyostelium*, p. 157-197. *In* A. A. Moscona and A. Monroy [ed.] Current topics in developmental biology, vol. III. Academic Press, New York.

*Since the last paragraphs were written, Bonner's laboratory has announced an exciting new discovery, the isolation and identification of a compound from the bacterium *E. coli* that has high attracting activity for the amoebae of *Dictyostelium discoideum*. This attractant is cyclic 3', 5'-adenosine monophosphate. The related species *Polysphondylium pallidum* already has been shown to synthesize it. *D. discoideum* is known to produce an attractant, but it has not yet been identified, apparently due to the secretion by the amoebae of large amounts of a phosphodiesterase that specifically breaks down cyclic 3', 5'-AMP acrasin to 5'-AMP. This discovery of the chemical nature of acrasin is surely an important step toward the future. Now that the chemotactic agent has been identified, one is in a position to study its specific effects on reactive cells. There are a number of questions that immediately come to mind: how does cyclic 3', 5'-AMP mediate chemotaxis, how does it make cells more adhesive, how does it supress center formation, and how is cyclic AMP itself controlled?

Harris, H. 1961. Chemotaxis. Exptl. Cell Res. Suppl. **8**:199–208.

Konijn, T. M., D. S. Barkley, Y. Y. Chang and J. T. Bonner. 1968. Cyclic AMP: a naturally occurring acrasin in the cellular slime molds. Am. Naturalist. **102**:225–233.

Miller, R. L. 1963. Chemotaxis during fertilization in the hydroid *Campanularia*. J. Exptl. Zool. **162**:23–44.

Oldfield, F. 1963. Orientation behavior of chick leukocytes in tissue culture and their interactions with fibroblasts. Exptl. Cell Res. **30**:125–138.

Rosen, W. G. 1962. Cellular chemotropism and chemotaxis. Quart. Rev. Biol. **37**:242–259.

Rothschild, L. 1956. Fertilization. John Wiley & Sons, New York. 170 p.

Shaffer, B. M. 1962. The Acrasina, p. 109–182. *In* M. Abercrombie and J. Brachet [ed.] Advances in morphogenesis, vol. II. Academic Press, New York.

Shaffer, B. M. 1964. Intracellular movement and locomotion of cellular slime-mold amoebas, p. 387–405. *In* R. D. Allen and N. Kamiya [ed.] Primitive motile systems in cell biology. Academic Press, New York.

Sussman, M. 1958. A developmental analysis of cellular slime-mold aggregation, p. 264–317. *In* W. D. McElroy and B. Glass [ed.] The chemical basis of development. The Johns Hopkins Press, Baltimore.

Twitty, V. C. 1966. Of salamanders and scientists. Freeman, San Francisco. 178 p. (especially Chapter IV).

Twitty, V. C. and M. C. Niu. 1954. The motivation of cell migration, studied by isolation of embryonic pigment cells singly and in small groups in vitro. J. Exptl. Zool. **125**:541–574.

FOUR

The Structural Basis of Cell Adhesion

Most tissue cells seem to be topographically stable in situ. Yet, if they are dissociated and mixed with other kinds of cells, they show a capacity for extensive changes in shape and even of translocation. Their immobility in situ therefore cannot be due to a lack of ability to move. They are held in place by adhesive forces. Selective adhesion of migratory cells like the neural crest and primordial germ cells during normal development and of cells of like origin in mixed cell aggregates (see Ch. 7) has led many to believe that these cell-to-cell adhesive forces may show tissue specificity. This in turn has stimulated widespread interest in the nature of the cell contacts.

Fine structure of cell contacts

As we begin our consideration of this subject, it must be emphasized that the nature of cell contacts is a highly speculative area of contemporary biology. Almost all physicochemical theories are based largely on models rather than on direct observations of the cell surface. In fact, we are not really in a position to define just what we mean when we refer to the "cell surface." There are, if you will, two general notions of the surfaces of cells, which might conveniently be called "dynamic" and "static." The dynamic surface is the one to which we have continually alluded. It is the cell surface we see in cultures of living cells and which in time-lapse films appears to be a changing, highly mobile, steady-state system. It may be relatively immobilized when another cell is encountered, as in contact inhibition; but when not in contact with other cells, as on glass, or in contact with other cells in a mixed aggregate of dissociated cells, it continues its mobility.

The static concept of the cell surface is based primarily on osmium-fixed material viewed in the electron microscope. The picture revealed is definite and remarkably consistent for cells, such as epithelial cells, that appear to be "in contact" at the limits of resolution of the light microscope. The cell membrane consists of a complex of two electron-dense lines with a less dense space between, the overall width being approximately 75 to 80 A (Fig. 4-1). This trilaminar membrane is the so-called unit membrane. There is a natural tendency to consider the electron-dense material as the plasma membrane. My generation and those before us have been conditioned by our histological training with fixed material to think of cells as composed of discrete membranes and organelles. Electron microscopy has reinforced this proclivity. But there are more cogent reasons for considering the unit membrane to be the outer membrane of the cell. There is much evidence that the layer at which surface tension and permeability properties are expressed is the absorbed monolayer on the outside of the bimolecular lipid leaflet, which Harvey and Danielli proposed for the structure of the plasma membrane. They believed that the absorbed layer is protein with a thickness of 15–20 A, since at that time no lipid surface was known that had the low surface tension properties typical of the plasmalemma. It is surely not coincidental that one electron-dense component of a unit membrane is about 25 A thick. Also, since phospholipids are highly osmiophilic, we would expect a layer rich in them to be electron dense after staining with osmium. It appears therefore that the unit membrane of the electron microscopist corresponds to the plasma membrane of the cell physiologist.

The electron microscope has revealed several different kinds of cell contacts. Since the number of these is small and they occur in similar histological situations in widely separated organisms, they appear to possess a kind of universality. Inasmuch as these contacts are described in detail in readily available histology and cytology texts, it is not necessary to devote much space to them here, except to lay a foundation for our discussion of the mechanisms of cell adhesions.

The closest cell contacts observed in the electron microscope are the so-called *tight junctions* or *zonulae occludentes* of Farquhar and Palade. In these junctions, the outer leaflets of the opposed unit membranes appear to be in contact or even fused (Fig. 4-2). In instances where the detailed structure

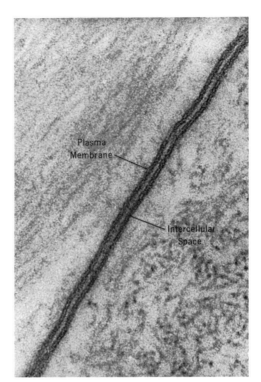

Fig. 4-1 Plasma membranes of two opposed glial cells of the annelid *Aphrodite* separated by a gap of ca. 150 A. ×260,000. (Fawcett. 1966. Atlas of fine structure. Saunders, Philadelphia. p. 448.)

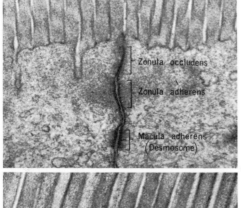

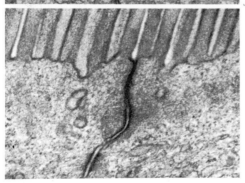

Fig. 4-2 Plasma membranes of two opposed intestinal epithelium cells from the hamster showing a tight junction (*zonula occludens*), a junction with a gap of 200 A between the opposed membranes (*zonula adhaerens*), and a desmosome (*macula adhaerens*). ×70,000. (Fawcett. 1966. Atlas of fine structure. Saunders, Philadelphia. p. 448.)

of the plasma membranes is not evident and the overall distance between the inner surfaces of the electron-dense material of the opposed membranes is of the order of 150 A, it may be concluded that tight junctions are probably present. Because tight junctions are found frequently in epithelial tissues, where the cells are tightly adherent, they are presumed to provide part of the basis for the adhesions. The most spectacular aspect of tight junctions comes from electrophysiological investigations. With the use of microelectrodes placed in cells, it has been found that cells joined by tight junctions invariably pass current readily from one to the other. They are in electrical communication. This of course means that they are also in chemical communication, at least in so far as inorganic ions are concerned. And there is evidence that larger molecules may also pass from cell to cell. Another important function of tight junctions is to block the passage of materials from one side of an epithelial cell sheet to the other between the cells. In this way the presence of tight junctions accounts in part for the reduced permeability of many epithelia. It has been shown, for example, that tight junctions prevent the flow of hemoglobin molecules between cells toward the lumen of a tubule.

Other intercellular junctions that appear to approximate tight junctions in some ways are *close junctions*. In these, the outer leaflets of the opposed membranes are very close to one another but not in contact. In this case, a clear separation of approximately 40–70 A may be detected (Fig. 13-4). Close junctions apparently do not occur frequently, and it is possible that some of them are modified tight junctions whose plasma membranes have been separated by fixation. Close junctions are just beginning to be studied, both at the fine-structural level and electrophysiologically. They too are regions of close cell adhesion and have been found in certain synapses. Recently, some presumed tight junctions have turned out to be very close junctions, with a gap of 20 A between the outer leaflets of the opposed plasma membranes. Such junctions were first shown in mouse liver and heart and most recently in the early *Fundulus* blastoderm (refer to Fig. 14-10). The gap between the unit membranes appears to be real, since it is penetrated by an electron-opaque preparation of lanthanum salts. When such a lanthanum-stained junction is cut tangentially, the cell membrane is shown to be composed of hexagonally packed structures. These "gap junctions" are of interest not only as probable sites of cell adhesion, but also because they may be areas of electrical communication between cells (see also p. 197).

While tight junctions are frequent, they are not extensive from the point of view of the area of cell surface which is involved. The bulk of the rest of the surface of an epithelial cell which abuts against its neighbors is separated from these neighbors by a distance of 100–200 A. Or, to put it more accurately, the less dense gap between the dense outer leaflets of the membranes of adhering cells is 100–200 A wide. It is referred to as a *100–200-A junction* or a *zonula adhaerens* (Fig. 4-2, Fig. 14-9). This type of cell relationship occurs very frequently and always appears to be associated with cohesion of the two cells, as in epithelia. This raises an interesting problem. If the gap is real and the cells are really separated by a distance as great as 200 A, how can we explain their adhesion to each other? It seems most unlikely that the gap is a fixation artifact. This type of junction is far too universal and appears under various conditions of fixation. Three possibilities are worthy of note. 1) The gap may be real, that is, filled with intercellular fluid containing largely low molecular weight compounds. This possibility is the most popular one and lies at the basis of an important theory of the mechanism of cell adhesions (see p. 84–85). 2) The gap is filled with high-molecular-weight intercellular material (or "cement") that serves to bind the cells together but has not been detected in electron micrographs because under the usual conditions of fixation and staining it is electron lucent. This is not at all improbable and will be considered at some length presently. One argument against this possibility, however, is that hemoglobin molecules and ferritin particles (100-A diameter) have been found to permeate between cells in the 100–200-A gap. If the ferritin perfusion is carried out after fixation, no such penetration occurs, suggesting that fixation either temporarily closes the gap or fills it with viscous material. 3) The gap is an artifact, in the sense that the unit membrane represents merely the electron-dense center of much thicker structures. If this were the case, the actual cell surfaces might be much closer than appears—perhaps 5–15 A.

As we shall see in considering various proposals for explaining cell contact and adhesion, this is a very critical question. We are faced with the problem of deciding where one cell leaves off and the other begins. Most workers in the field, however, choose the unit membrane, as visualized in electron micrographs, as the plasma membrane. A compelling reason for this is the fact that the unit membrane always appears to have approximately the same dimensions, no matter what its distance from the membrane of a neighboring cell. And this distance may vary from zero, in tight junctions, to several hundred angström units, when cells appear not to be in adhesive contact. Since, however, there is a genuine paucity of information when dealing with details of membrane structure at this level, it is wise to keep an open mind.

In some kinds of cells the only apparent structural basis for cell-to-cell adhesion are tight and close junctions and 100–200-A junctions, all of which tend to be zonular, extending like a belt around the cell surface. Some cells, however, are obviously more adhesive in certain discrete regions of the surface. If they are shrunk by immersion in a hypertonic medium or pulled apart mechanically, they tend to stick more firmly at certain points. When examined in the electron microscope these plaques of greater adhesiveness show a distinctive structure which stains densely with osmium tetroxide. These plaques are called *desmosomes* or *maculae adhaerentes*. As may be seen in Fig. 4-2, they are local differentiations of the unit membrane that in preparations of normal tissue are in exact apposition with those of the next cell. They are common in epithelia of adults or advanced embryos and especially abundant in stratified squamous epithelium. Desmosomes consist of two electron-dense symmetrical plaques on the opposing cell surfaces separated by a gap of 200–250 A. On each half-desmosome a thin layer of dense material coats the inner leaflet of the plasma membrane, and bundles of thin cytoplasmic filaments usually converge on and terminate in this dense layer. Often the intercellular space is bisected by a slender intermediate line. Their major and perhaps only function is in providing structural basis for cell-to-cell adhesion.

With this in mind, Overton examined cells of the chick blastoderm at various early stages of development in the electron microscope. Even though cells of the pellucid area of a primitive streak blastoderm are firmly adherent, they completely lack desmosomes. These structures are first observed at stages 5 to 7 (head process to first somite) and are more prominent and more numerous at stages 10 to 11 (thirteenth to nineteenth somites). During these stages, the facility with which blastoderms may be dissociated into suspension of single cells decreases roughly as the prominence and number of desmosomes increase. It would appear, therefore, that desmosomes do indeed play a role in cell-to-cell adhesion, but this role is supplementary to other mechanisms for which there are no obvious structural differentiations (such as *zonula occludens* and *zonula adhaerens*). Overton went on to study the fine structure of reaggregating cells and found that between 4 and 13 hours after reassociation desmosomes gradually reappear, following a sequence similar to their normal development. Although unaligned half-desmosomes were noted, most half-desmosomes were in register with those of the opposed cell. Since reestablishment of normal structural relations between cells paralleled their reassociation, it seems reasonable to suspect that the desmosomes again are causally implicated. Although desmosomes appear to be fixed structures that bind cells together in a relatively stable way, this is not necessarily so. Desmosomal attachments between cells of the basal and prickle layers of the epidermis, for example, cannot be fixed structures, because labeling studies with tritiated thymidine show that basal cells move constantly into the upper layers as individuals. During this migration desmosomes continually form and reform.

Tight junctions and desmosomes appear to be limited to vertebrate tissues. Their place in invertebrates is taken by so-called *septate desmosomes* (Fig. 4-3). These resemble tight junctions in that they tend to be found in epithelia, are *zonula*, and involve only a small extent of the plasma membranes of the opposed cells. In some other respects, however, they are quite different. The unit membranes are separated by a gap of approximately 200 A, which is crossed by a number of septa. These septa appear to be extensions of the outer leaflets of the opposed unit membranes and attach the two cells to each other. There are no intracellular deposits backing up septate desmosomes (as in the vertebrate desmosome), and no intracellular fibrils converge on the site. Because of the obvious structural connections the septa make with the two opposed cells, septate desmosomes are universally considered as adhesive or attachment devises. They possess

Fig. 4-3   Septate desmosome connecting epithelial cells lining the water vascular system of the sea urchin *Arbacia punctulata*. (Courtesy of J.-P. Revel and E. Hay. Harvard Medical School, Boston.)

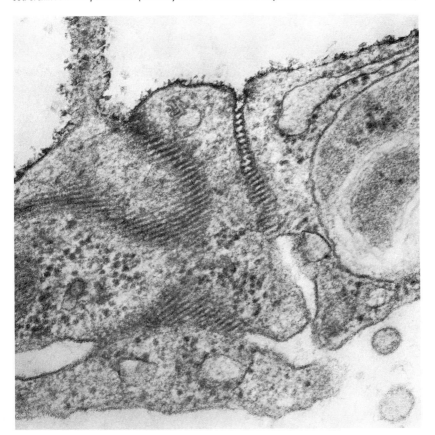

one more important similarity with tight junctions. They are apparently sites of low resistance which facilitate electric communication. Invertebrate cells with septate desmosomes are electrically connected; those without are not.

In some instances, cells which appear to be closely adherent are revealed by the electron microscope to have highly folded membranes which interdigitate with the folded membranes of adjacent cells (Fig. 14-11). The membranes in such cases are often separated by the classical 100–200-A gap. Apparently these interdigitations aid cell adhesion. Their appearance suggests this as a likely function and in the proximal regions where the two cell membranes are not bound together at the septate desmosome the membranes often separate widely.

Although intercellular materials or "cell cements" have long been thought essential for building cells together into tissues, the electron microscope initially provided little support for this belief. With osmium and potassium permanganate fixation, even after pretreatment with gluteraldehyde, most cells possess no electron-dense material beyond the outer leaflet of the plasma membrane. With such treatment, for example, the famous 100–200-A gap almost invariably appears devoid of materials. Besides, the need for an intercellular cement was largely vitiated by the discovery of admirable attachment devices, such as tight junctions, desmosomes, and septate desmosomes. To be sure, a number of cell types were shown to possess a surface coating of materials, but only on exposed terminal surfaces where the cells were not in contact with other cells. The brush borders of intestinal epithelial cells, for example, are covered with a dense tangle of extracellular filaments, which are attached to the outer leaflet of the plasma membrane. But in the intercellular zone such materials seemed to be largely lacking.

Very recently, however, the use of special staining techniques has reopened the question in an impressive way. Periodic acid-silver methenamine, a fairly specific technique for glycoprotein detection, was applied to rat tissues, which were then examined under the electron microscope (Fig. 4-4). It was found that nearly all cells are coated with a thin layer of stained material. This layer extends between cells everywhere, except where the plasma membranes of two cells fuse to form tight junctions. This suggests that the stained layer is intercellular material, located outside the plasma membrane. The material presumably contains glycoproteins, but the presence of other substances is not excluded. Another method has been to fix tissues in the presence of lanthanum ions. Chick embryonic tissues treated in this way show the presence of an electron-opaque layer, about 50 A thick, external to the plasma membrane and continuous with its outermost leaflet. It is present wherever there is space between cells, but not where the membranes are in contact in a tight junc-

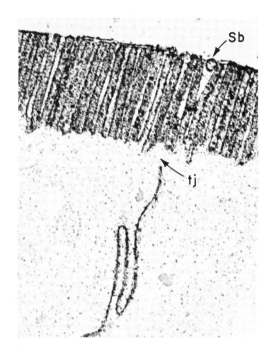

Fig. 4-4 Striated border (Sb) of epithelial cells of the small intestine of the rat stained with periodic acid-silver methenamine. Note the stained material in the gap between the cells except where the tight junction (tj) is located ($\times 25{,}500$). (Courtesy of A. Rambourg and C. P. Leblond. After Rambourg and Leblond. 1967. J. Cell Biol. **32**:27.)

tion. It can be removed with phospholipase C but not with trypsin, pronase, or EDTA, which are efficient at dissociating cells. When it is removed, the plasma membrane remains intact. All of this suggests that this lanthanum-staining material, like that stained with periodic acid-silver methenamine, is intercellular material, located outside the cell membrane. Because of their intimate association with the cell membrane, these materials may serve functionally to extend the membrane and thus in effect bring cells closer together. In this manner, they could promote cell-to-cell adhesion. The general subject of intercellular materials is a large one, with a long history, many ramifications, and profound implications for the physicochemical mechanisms of cell adhesions. For this reason it will be treated more extensively in the next chapter.

Cytoplasmic bridges

Two additional means of constructing tissues are worthy of note, even though they are of rather limited occurrence. These do not involve the opposition of cell surfaces, but partial or total fusion of cells. One of these has found its clearest description in electron micrographs. These are the so-called cytoplasmic bridges. They are generally remnants of incomplete cell divisions, and as long as they persist the two or more cells so united are indeed closely bound together. Some of the best examples are to be found among spermatocytes, where it is thought that these organic cytoplasmic connections promote synchrony of meiosis in the attached partners. They are also often found connecting cleaving blastomeres during early embryonic development, when cell divisions are highly synchronous. Whatever their importance may be for the mitotic mechanism, cytoplasmic bridges are of little interest to us in our consideration of cell adhesion. They have been found in only a few kinds of cells, and the great majority of tissue cells are able to adhere very well in their complete absence.

Cell fusion

The ultimate in cell contact is cell fusion. There are not many instances known of the fusion of cells under physiological conditions, but it so happens that the best known, that of the sperm and the egg, is also the best analyzed from the point of view of fine structure. The Colwins studied sperm and egg fusion in the annelid *Hydroides* and the protochordate *Saccloglossus* and obtained a remarkable series of electron micrographs that show the process in detail (Fig. 4-5). The membrane of the sperm acrosome first fuses with the plasma membrane of the sperm. Then, as an extension of the sperm plasma membrane, the acrosomal membrane makes contact and fuses with the plasma membrane of the egg. The sperm and the egg are thus joined to form a single cell before "sperm penetration"; and the zygote has a plasma membrane which is a mosaic, composed of the plasma membrane of the unfertilized egg, the acrosomal membrane of the sperm, and the plasma membrane of the sperm. These observations are most suggestive. Not only does an internal membrane fuse with the surface membrane of a cell, as if they were made of similar stuff; it even fuses with the plasma membrane of another and entirely different cell derived from another individual. Is it possible that cell membranes do not differ qualitatively?

The most complete result of cell fusions is a syncytium, a condition in which a mass of cytoplasm is populated by many nuclei, with no cell boundaries separating the nuclei. The most well-known syncytium is the skeletal muscle fiber, in which there may be hundreds of nuclei. A series of elegant experiments, in which embryonic myoblasts were labeled with tritiated thymidine and observed in clonal culture, have shown that muscle fibers are formed in vitro by the fusion of originally discrete cells, with the complete disappearance of cell membranes. Proof that muscle fibers form in the same way in vivo awaited a rather extraordinary investigation in which early embryos of mice differing in the genes for isocitrate dehydrogenase were fused in vitro. These mosaic embryos were then transplanted

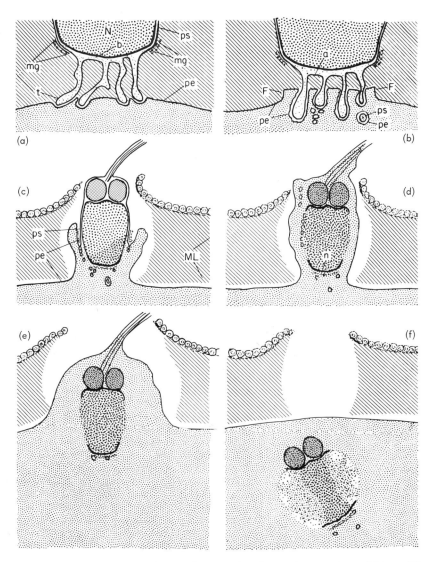

Fig. 4-5 Diagrams of stages in fusion and incorporation of spermatozoon and egg of the annelid, *Hydroides hexagonus*. (a, b) acrosomal tubules of sperm head indent egg, but egg plasma membrane intervenes between egg cytoplasm and sperm plasma membrane; (c) fusion of egg plasma membrane with sperm plasma membrane, vesiculation near presumed site of fusion, acrosomal tubules no longer visible, egg cytoplasm now in direct contact with apical part of nucleus; (d) within their common plasma membrane egg cytoplasm surrounds sperm structures, but fused gametes still have profile of both egg and spermatozoon; (e, f) sperm structures move more deeply into egg cytoplasm, mitochondrial and apical parts of nuclear envelope remain still visible but peripheral part of nucleus becomes diffuse, fertilization cone recedes. mg, granules, or remnant, of material of intermediate zone of acrosome; t, acrosomal tubule; ps, plasma membrane of spermatozoon; pe, plasma membrane of egg; F, fertilization cone; a, that part of sperm plasma membrane which was formerly acrosomal membrane; ML, middle layer of vitelline membrane; n, nuclear envelope. (Colwin and Colwin. 1961. Cellular membranes in development. Academic Press, N.Y. p. 233.)

to the uteri of surrogate mothers and eventually raised to adulthood. Each allelic variant produces only a single band on electrophoresis, and the heterozygote produces the two parental enzymes and a third or hybrid enzyme. If the skeletal muscle fibers of these mosaic animals form from the multiplication of the nuclei of single myoblasts without accompanying cytoplasmic division, the fibers would contain one or the other of the parental enzymes but not the hybrid enzyme. On the other hand, if the muscle fibers form by the fusion of hundreds of separate myoblasts, the two genetic types, both parental enzymes, *and* the hybrid enzyme would be present in the syncytium. The hybrid enzyme was present in the skeletal muscle of all mosaic animals, demonstrating conclusively the in vivo origin of this syncytium by myoblast fusion. Analysis of a variety of other tissues from these mice disclosed no hybrid enzyme, indicating a lack of cell fusion.

This last discovery is of much interest because of the numerous instances that have been detected during the last few years of fusion of various nonmuscle cells in culture. It appears as if there is something peculiar to the culture conditions and not present in the organism which promotes these fusions. In any case, these fusions in culture differ from the fusion of myoblasts in two significant ways. Apparently only two cells fuse at a time, and the fusion involves the nuclei as well as the cytoplasm. In skeletal muscle fibers individual nuclei remain intact and separate from each other, though sharing a common cytoplasm. None of these fusions of somatic cells has been studied with the electron microscope. It would be most interesting to compare the manner of fusion of these cells with that of the sperm with the egg, as in the investigations of the Colwins.

Another syncytium whose physiological activity is also spectacular is the yolk syncytium or periblast of meroblastic eggs. This nucleated sheet spreads over the yolk eventually to encompass it entirely during the process of epiboly. Because of its position between the blastoderm and the yolk, it serves as the intermediary between the blastoderm and its nutrient supply. It is of little interest in the present context, however, because it is not formed by the fusion of cells but rather by the invasion of uncleaved cytoplasm by nuclei of blastomeres whose boundaries are not yet complete.

These syncytia, and others not described, are clearly one means of tissue formation and for this reason deserve comment at this point. However, like cytoplasmic bridges, they are not really germane to the problem before us—viz., how do cells, which may be readily and viably separated by chemical dissociating agents and which adhere in tissues and cell aggregates as discrete cells, adhere selectively to each other?

SELECTED REFERENCES

Colwin, A. L. and L. H. Colwin. 1964. Role of the gamete membranes in fertilization, p. 233–279. *In* M. Locke [ed.] Cellular membranes in development. Academic Press, New York.

Curtis, A. S. G. 1962. Cell contact and adhesion. Biol. Rev. Cambridge Phil. Soc. **37**:82–129.

Ephrussi, B. and M. C. Weiss. 1967. Regulation of the cell cycle in mammalian cells: Inferences and speculations based on observations of interspecific somatic hybrids. *In* M. Locke [ed.] Control mechanisms in developmental processes. Develop. Biol. Suppl. **1**:136–169.

Farquhar, M. G. and G. E. Palade. 1963. Junctional complexes in various epithelia. J. Cell Biol. **17**:375–412.

Fawcett, D. W. 1961. Intercellular bridges. Exptl. Cell Res. Suppl. **8**:174–187.

Fawcett, D. W. 1966. An atlas of fine structure. The cell: its organelles and inclusions. Saunders, Philadelphia, 448 p.

Konigsberg, I. R. 1963. A clonal analysis of myogenesis. Science. **140**:1273–1284.

Lesseps, R. J. 1967. The removal by phospholipase C of a layer of lanthanum-staining material external to the cell membrane in embryonic chick cells. J. Cell Biol. **34**:173–183.

Mintz, Beatrice. 1967. Normal mammalian muscle differentiation and gene control of isocitrate dehydrogenase synthesis. Proc. Natl. Acad. Sci. U.S. **58**:592–598.

Rambourg, A. and C. P. Leblond. 1967. Electron microscopic observations on the carbohydrate rich cell coat present at the surface of cells in the rat. J. Cell Biol. **32**:27–53.

Robertson, J. D. 1964. Unit membranes: a review with recent studies of experimental alterations and a new subunit structure in synoptic membranes, p. 1–81. *In* M. Locke [ed.] Cellular membranes in development. Academic Press, New York.

Stockdale, F. E. and H. Holtzer. 1961. DNA synthesis and myogenesis. Exptl. Cell Res. **24**:508–520.

Wolpert, L. and E. H. Mercer. 1963. An electron microscopic study of the development of the sea urchin embryo and its radial polarity. Exptl. Cell Res. **30**:280–300.

Wood, R. L. 1959. Intercellular attachment in the epithelium of *Hydra* as revealed by electron microscopy. J. Biophys. Biochem. Cytol. **6**:343–352.

FIVE

The Measurement of Cell Adhesion

Since measurement is basic to precise definition, a means of measuring the adhesiveness of cells is basic to a definition of cell adhesion. Attempts to measure the adhesiveness of cells have fallen into two general categories: physical separation and reaggregation kinetics. Procedures for separating cells physically seem to have been of two kinds: 1) measuring the force required to separate a cell from another cell or a population of cells from a surface, and 2) applying a standard force and determining the proportion of the cell population that is separated from each other or removed from a surface.

There have been several attempts to measure adhesiveness by measuring the force required for separation. Cells have been detached from glass tubing with a stream of fluid. Centrifugation has also been used. Another method has been to measure adhesiveness directly by pulling cells from one another with glass microneedles. Coman, who has been a pioneer in this kind of investigation, judged degree of adhesiveness by the force required to separate cells in culture. He estimated force by the curvature produced in the moving needle and by this means compared adhesiveness between normal cells (lip and uterine cervix) and neoplastic cells (both benign and malignant). He found that the neoplastic cells were much more readily separated than those from normal tissues and concluded that they are less adhesive. It is of interest in this connection to recall that cancer cells are not contact inhibited by normal cells. Could this be due to lesser adhesiveness of cancer cells?

Since cells must adhere to the substratum in order to flatten and spread on it, the degree of adhesiveness of the cell surface might determine the degree of flattening. Conversely, the degree of flattening could give a rough indication of degree of adhesiveness. If several surfaces are available, as when a cell is surrounded by other cells, an adhesive cell will tend to adhere to all of them, conforming to their surfaces. This results in the polyhedral cell contours of closely packed tissues. But if only one surface is available and cells are more adherent to it than to each other, they will tend to flatten on this surface; the larger the adhering area, the greater degree of flattening. It has been shown recently that there is a direct correlation between degree of flattening and degree of adhesiveness, as judged by the proportion of cells adhering to a substratum after being subjected to turbulence in the medium. This bolsters the notion that degree of flattening may be used as an approximate measure of degree of adhesiveness. However, as pointed out in Ch. 2, flattening might also be caused by several other factors. Hence, even though it is sometimes useful as a working hypothesis to consider degree of flattening as indicating degree of adhesiveness, such may turn out to be invalid.

With the use of a standardized force to separate cells, an early attempt by Coman and his co-workers gave a result that corroborates the conclusion reached by pulling cells apart with needles. Intact tissues were agitated in a standardized way in fluid medium by means of a mechanical shaker. When counts were made of both the remaining pieces of tissue and of the cells torn loose, it was found that more cells were liberated from neoplastic than from normal tissues. If it be objected that the in vitro conditions employed were unphysiological, a ready answer would be that leukocytes and macrophages, which are freely mobile through tissues, are similarly not readily adherent to each other in vitro. At this juncture, therefore, the conclusions appear to be justified.

The last few years have seen the introduction of more sophisticated methods, in which attempts are made to quantify accurately the amount of shearing force required to dislodge cells adherent to surfaces of various materials. L. Weiss tried by rotating a metal disc at a standard speed, in a medium of known viscosity, at a known distance from cells adherent to a particular inert substratum. After rotation, the strength of cell adhesion to the substratum could be gauged by the proportion of cells dislodged.

It might seem to some that we now have sufficiently refined methods for estimating the surface adhesiveness of cells and are therefore in a position to determine with some accuracy whether postulated differences in adhesiveness in vivo, in mixed aggregates, and in other complex systems actually exist. Unhappily, such optimism is unjustified. All of these techniques make the assumption that the adhesive force holding two adherents together is invariably the same as the force required to pull them apart. This is not necessarily so. L. Weiss has pointed out that if the cohesive strength of the cell surface is less than the adhesive strength of the cell-substratum interface (joint), then on distraction rupture will not occur in the joint between the two but will occur within the cell surface, and part of it will be left attached to the substratum. If, on the other hand, the cohesive strength of the substratum is less than that of the joint, the substratum will rupture on distraction, and part of it will be removed with the cell. This will give a measurement of substratum strength.

In support of these ideas, L. Weiss presents the results of some of his own studies. When fibroblasts are detached from a glass substratum by shearing forces, the glass surface is altered in such a way that trypsinized cells now adhere to it more strongly in serum-free media than to untreated glass. This alteration of the glass surface does not occur when the cells are suspended in hanging drops away from the glass. The increased adhesiveness of the glass for the cells is eliminated by trypsin treatment. It would therefore appear that when fibroblasts are pulled away from their glass substratum they leave some proteinaceous material behind. In another study, it was shown by a rather ingenious technique that when two other strains of cells are pulled away from glass, they leave antigenic material behind. Species-specific erythrocyte antibodies combine with this antigenic material as shown by the agglutination to it of erythrocytes of the same species. Fluorescein-labeled antibodies also combine with it. Although there can be no reasonable doubt that the cells have left something behind in these experiments, both the nature of the material and its precise origin are left in doubt. It has not been shown whether it represents fragments of ruptured cell membrane as postulated, or extracellular material, or exudate that has oozed from the cells under culture conditions. Again, we are left in an unsettled state. There is indeed a possibility that the permanence of cell adhesions depends less on the strength of the adhesion than on the strength of the adhesive membranes. Yet, in spite of the ingenuity devoted to the experiments, the evidence which has thus far been provided is inconclusive.

This is not the only difficulty. Even if the force required to pull cells apart is the same as that holding them together, measurement of the force alone would not be sufficient. The work required to pull cells apart or a cell away from its substratum can be determined only by integrating the force over the distance through which the force is exerted. Steinberg has emphasized that the amount of force required would vary with the manner in which a cell is pulled from its substratum (Fig. 5-1). A very great force would be required to pull it off vertically, all at once. If, however, a cell were peeled off its substratum, the amount of force required would be very small at first, increasing to a peak as the diagonal is reached

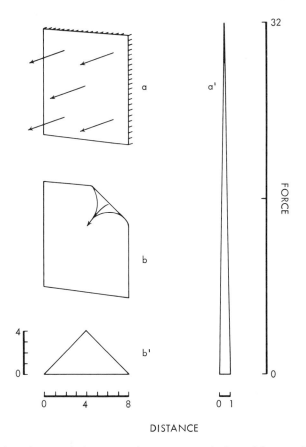

Fig. 5-1 Relation between the force required to separate two bodies and the strength with which they cohere. A far greater force is required to pull a square of material from a surface all at once (a, a') than to peel it off (b, b'). In a-a' the large force is exerted over a small distance. In b-b' the small force is exerted over a large distance. The work (force × distance) is the same in both cases. It, rather than the peak force exerted, is the measure of the strength of cohesion. (Steinberg. 1964. Cell membranes in development. Academic Press, N.Y. p. 321.)

and declining to zero as the last adhesive bonds are ruptured. At no time would the force be nearly as great as when a cell is pulled off all at once. Yet the amount of work done in each instance would be the same, since in the latter case the force is applied over a very short distance. It is evident that simple determination of the force required to pull cells off a substratum is not enough. If cells pull off at once, a large force applied for a short time would be necessary. If cells peel off gradually, a small force applied over a long time would suffice. Careful determination of the force and the time required for distraction, along with observation of the manner of distraction, should enable one to make more meaningful determinations. Use of the centrifuge microscope, for example, could provide such information. But there are still other problems. It appears that rheological properties of the cell surface, such as rigidity and plasticity, may be altered by the rate of application of a distractive force. Since these may in turn influence the force required for distraction, the rate of application of the distractive force could of itself influence the amount of force required. In this regard, it is of interest that L. Weiss found that the measured strength of adhesion varied with the rate of application of shearing forces.

One even encounters difficulties in the interpretation of data on the percentage of a cell population cohering with each other or adhering to an inanimate substratum during a given time interval. There is evidence that the probability of initiating an adhesion is a separate function from the strength of such an adhesion after it has been initiated. For example, an increase in temperature increases the possibility of initiating an adhesion without noticeably affecting the stability of the resulting attachments. Thus it is quite possible that cells that are weakly adhesive may nonetheless adhere to each other or to an inert substratum more readily than those that are potentially strongly adhesive.

Finally, we have little assurance that cellular adhesion to glass, plastic, or other inert substrata relates in any direct way to adhesion of cells to other cells. To be sure, Curtis has shown that a 100–200-A gap separates certain cells from a glass substratum; but knowing what we do about the highly structured basis for cell-to-cell adhesion in some instances, it would be surprising indeed if adhesion to an inert substratum and adhesion to cells were always similar. For this reason a more direct approach is necessary, in which the adhesion of cells to cells is studied.

The second general category of measuring cell adhesions—reaggregation kinetics—deals with adhesions between cells and in particular with the formation of these adhesions, rather than their disruption. With this method, a suspension of cells is agitated so that cells collide and adhere. The probability that permanent adhesions will form between cells in suspension is assumed to be directly related to their mutual adhesive stabilities. The size and number of resulting aggregates, the rate of disappearance of single cells, and the rate of appearance of cell clumps have been assumed to indicate the degree of adhesiveness between the cells in a suspension. A difficulty common to these techniques, however, is that they do not distinguish between two alternatives. Observed differences in adhesiveness between cells may reflect intrinsic adhesive properties characteristic of a particular cell type or temporal artifacts caused by recovery from damage to the cell surface by the agents used to dissociate the cells.

In any technique using reaggregation of cells, the net number of adhesions between cells will be the result of the number of collisions that occur during a given time multiplied by a certain probability that any collision will result in a stable adhesion. The number of collisions will of course be influenced by a number of extraneous factors such as concentration of cells, effective diameters, viscosity of the medium, speed of agitation, and density of cells. If characteristic differences in cell adhesion are to be detected, these extraneous factors must be controlled. Roth and Weston accomplished this and eliminated temporal artifacts recently by circulating unlabeled cell aggregates as collecting fragments in aliquots of a cell suspension labeled with tritiated thymidine. Results were acquired in terms of the number of labeled cells collected by these unlabeled aggregates, as determined in radioautographs. If all extraneous factors are the same for the cell types being compared, adhesions between isotypic and heterotypic combinations can be compared. In this way one can determine the relative adhesiveness of various cell types for each other. Unlabeled aggregates of liver and neural retina from a seven-day chick embryo were circulated in suspensions of neural retina and liver cells labeled with tritiated thymidine. It was found (Fig. 5-2) that aggregates in isotypic combinations (aggregates of liver plus labeled liver cells and aggregates of neural retina plus labeled neural retina cells) collected many more labeled cells than did aggregates in heterotypic combinations (aggregates and suspensions of different cell types). These results demonstrate that: 1) striking differ-

Fig. 5-2 Radioautographs of chick neural retina (a) and liver aggregates (b), which have circulated in a labeled neural retina suspension 4×10^5 cells/ml for 5 hr. Labeled cells adhere primarily to isotypic aggregates. Scale = 50 μ. (Courtesy of Stephen Roth. After Roth and Weston. 1967. Proc. Natl. Acad. Sci. **58**:974.)

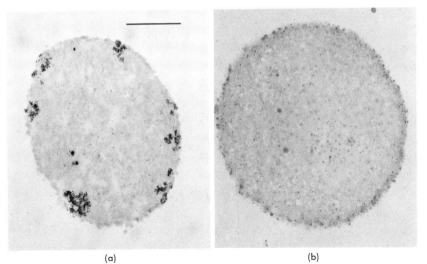

(a) (b)

ences in cell-to-cell adhesiveness exist; 2) these differences in cell adhesiveness are type specific, with greater stability in isotypic associations than in heterotypic associations; and 3) this specificity of adhesion exists before they get into aggregates, i.e., it is intrinsic to the cells.

With the technique of Roth and Weston we have a direct means of determining whether the degree of adhesiveness of cells to glass, plastic, and other inanimate substrata bears any relation to their adhesion to other cells. The pressing need of such studies cannot be overemphasized. A large proportion of our thinking about differences in cellular adhesiveness is based on the adhesion of cells to inanimate substrata.

We are now also in a better position to advance an operational definition of intercellular adhesion. The definition of Curtis seems the most useful yet proposed. He defines intercellular adhesion as the probability that two cells will form and maintain an adhesion.

The reader may wonder why a chapter has been devoted to a body of results that is largely inconclusive. It has been necessary in order to place in proper perspective the various theories of the mechanisms of cell adhesion and association that we are about to discuss and to emphasize the pressing need for both hard thinking and clever experiments. Often theories are proposed to explain differences in degree of adhesion where the measurement of the degree of adhesion has not yet given unequivocal results. Even before we move on to consider various theories of cell adhesion you will no doubt have already guessed that we are in no position at the moment to decide among them. Nevertheless, they deserve serious attention. Science must move on many fronts. Careful observation of the ways cells behave may suggest, support, or eliminate hypotheses which try to explain this behavior. And speculations based largely on simpler systems may suggest the need for particular observations that no one until then had thought of making.

SELECTED REFERENCES

Coman, D. R. 1944. Decreased mutual adhesiveness, a property of cells from squamous cell carcinomas. Cancer Res. **4:**625–629.

Curtis, A. S. G. 1964. The adhesion of cells to glass. A study by interference reflection microscopy. J. Cell Biol. **19:**199–215.

Roth, S. A. and J. A. Weston. 1967. The measurement of intercellular adhesion. Proc. Natl. Acad. Sci. U.S. **58:**974–980.

Steinberg, M. S. 1964. The problem of adhesive selectivity in cellular interactions, p. 321–366. *In* M. Locke [ed.] Cellular membranes in development. Academic Press, New York.

Weiss, L. 1962. Cell movement and cell surfaces: a working hypothesis. J. Theoret. Biol. **2**:236–250.

Weiss, L. and P. J. Lachmann. 1964. The origin of an antigenic zone surrounding Hela cells cultured on glass. Exptl. Cell Res. **36**:86–91.

SIX

Mechanisms of Cell Adhesion

In considering mechanisms of cellular adhesion the investigator possesses a high degree of freedom. Unhampered by an array of hard facts concerning cell contact behavior, he is free to speculate almost at will. This has resulted in a plethora of hypotheses which seek to explain on physicochemical grounds the forces which hold cells together. In this chapter we shall consider the most prominent of these. Each hypothesis has usually been proposed as the exclusive means whereby cells adhere. It seems possible, however, that in some cases a number of mechanisms may act in concert.

Molecular complementarity hypothesis

On theoretical grounds the closest kind of adhesion that can occur between cells involves direct chemical bonds between their surfaces. Examples of linkages that can occur are hydrogen, ester $\left(-\overset{\overset{\displaystyle O}{\|}}{C}-OR\right)$, thiol $(-S-S-)$, imine bonds $\left(-\overset{\overset{\displaystyle H}{|}}{C}=N-\right)$, and amide bonds $\left(-\overset{\overset{\displaystyle O}{\|}}{C}-\overset{\overset{\displaystyle H}{|}}{N}-\right)$. These would cause strong adhesions. Specific adhesion could be imagined as resulting from the patterned complementation of active groups (like —SH, —OH, and —NH$_2$) between the two cell surfaces or by ion pairing (the patterned complementation of anions of one cell surface with cations on the other cell surface). The main ions involved would be carboxyl $\left(\begin{array}{c}\text{O}\\\|\\-\text{C}-\text{O}^-\end{array}\right)$ and amine groups $\left(\begin{array}{c}\text{H}\\|\\-\text{N}^+-\text{H}\\|\\\text{H}\end{array}\right)$ of the cell-membrane proteins and phosphoryl ions $\left(\begin{array}{c}\text{O}\\\|\\-\text{O}-\text{P}-\text{OH}\\|\\\text{O}^-\end{array}\right)$ of the cell-membrane phospholipids.

The first attempt to explain type-specific cell adhesion by some form of molecular complementarity was that of Tyler and of Weiss over twenty years ago. They independently proposed that each molecular contact is specific and like an antigen-antibody reaction. The cell surface was considered to contain two types of reactive groupings, so that when two like surfaces came together bonds were formed. They further suggested that qualitative or quantitative alterations in the adhesiveness of cells may be a reflection of changes in the antigenicity of their surfaces. Surface-specific antigens have been demonstrated in many different kinds of living cells, from bacteria to mammals. Spiegel attempted to test the hypothesis by exposing sponge cells to species-specific antisera made against them. These experiments excited quite a lot of interest when he was able to reversibly block initial adhesions between cells of one species with antibody made against cells of that species. Gregg demonstrated that homologous antiserum will also block agglutination of cells of the cellular slime mold *Dictyostelium discoideum* and then went on to show by the use of agglutination techniques that new surface antigens are formed in cultures of slime molds as the time for aggregation approaches. It is possible that the new antigens are concerned with adhesion. He also showed that cells with antigenically dissimilar surfaces would not perform coordinated morphogenetic movements.

A variant of this theory is that particular large molecules form a cement between cells. In agglutinated blood cells and certain bacteria agglutinated with antibody the surfaces may be bound together with antibodies. In this way erythrocytes have been packed together so tightly that they resemble a cohesive tissue.

There are at least three difficulties with the specific-combining-group hypothesis. It is of course probable that cells have combining groups at their surfaces; the question, however, is whether cells adhere to one another by virtue of these complementary sites. In the first place, these theories demand that the surfaces come within 2–3 A units of each other. This raises a problem. In most cell adhesions the distance between the electron-dense unit membranes of cell surfaces is of the order of 100–200 A units (Fig. 4-1). If the gap contains cementing materials that contain specific binding groups, there is no difficulty. Secondly, it has often been observed that cells will adhere to chemically inert materials like gold, tin, and tantalum. How can this observation be reconciled with a hypothesis requiring chemical combination for adhesion? Finally, although blocking adhesion with homologous antisera is consistent with the hypothesis, it does not necessarily support it. P. Weiss, one of the originators of the theory, warns us (1958) that interference with cell aggregation by species-specific antibodies "can only prove, at best, that surface molecules which are involved in cell adhesion also have antigenic properties, but not that these very same properties are instrumental in the normal coupling of cells." Spiegel himself has pointed out that his results could also be interpreted as meaning that antibodies react with groups near nonspecific aggregative groups and prevent adhesion by some sort of steric hindrance, without playing a direct role in the adhesion process. Moreover, it is not certain that antibodies in the experiments act at the cell surface. They could have their inhibitory effects by penetrating inside the cell.

The support given to the idea that antibodies may form a cement between cells from the appearance of agglutinated erythrocytes is rendered less impressive by a closer look in the electron microscope (with ordinary fixation). Dense staining was found between the erythrocytes, unlike the weak staining between cells in normal adhesion.

The most impressive support for the molecular complementarity hypothesis comes from study of the mating of yeast cells. When cell suspensions of two mating types of the yeast, *Hansenula wingei*, are mixed, massive agglutination occurs due to the strong adhesive forces between cells of opposite mating types. The agglutination is highly specific. Neither mating type agglutinates with itself or with the diploid hybrid between the two. Recent work in Brock's laboratory has shown that a cell-surface agglutination factor may be isolated from yeasts of one mating type by enzymatic digestion of whole cells, which specifically agglutinates cells of the opposite mating type. No cell agglutinating factor has yet been isolated from the opposite mating type; however, its cells have yielded a component that specifically inhibits the agglutinating activity of the agglutinin of its opposite. These factors have been shown to be glycoproteins of low molecular weight. Clearly, they are complementary macromolecules that neutralize each other as do antibodies and antigens. Since they are derived from extracts of whole cells it is not yet certain that they come from the cell surface. But because a) their activity is clearly at the cell surface and b) their combining specificity corresponds to that of the whole cells, the tentative conclusion seems justified that they are indeed surface factors.

Calcium-bridge hypothesis

A second theory for explaining cell adhesion assumes that the molecular contacts are alike for all cell types, or, if specificity of contact is required, that the pattern or arrangement of contacts may be specific for each cell type. In this way, only like cells would be able to form a sufficient number of bonds among one another to remain adherent. This type of theory is an outgrowth of a long-established fact. Herbst first demonstrated around the turn of the century that if cleaving sea urchin eggs are placed in Ca^{++}-free seawater, one can easily shake the blastomeres apart after a few minutes. Since then it has been shown time and again that cells of many kinds of embryos may be dissociated if left in Ca^{++}-free media. Presumably the efficacy of the chelating agent, EDTA, for dissociating cells is due in particular to its ability to bind calcium. Addition of Ca^{++} ions to a medium deficient in that ion will promote adhesion. As may be guessed, calcium does not function alone in this regard. Other divalent cations, such as magnesium and strontium, are also implicated, though to a lesser extent. It has been proposed that each cell type has a particular pattern of anionic binding sites that are bound to those on an adjacent cell by bridges of divalent calcium ions. This theory requires a fairly rigid arrangement of anionic binding sites. Otherwise, patterns could be easily distorted and not remain specific.

Steinberg, who was once a champion of this "calcium-bridge theory," postulated the existence of patterns of exposed anionic groups with differing spacings and showed that this would permit cells with like patterns to form more cross-links and cells with unlike patterns fewer. Consequently, all cell types would adhere to some degree (as in fact is the case), but like cells would form more adhesions and thus cohere more strongly. There is much indirect evidence that calcium binds to the cell surface, and it has been demonstrated directly with ^{45}Ca. The shape of the titration curves of complexing of ^{45}Ca suggests that the main binding sites on the cell surface are carboxylic acid end groups.

There are some serious objections to the calcium-bridge theory as an explanation of specific adhesion. Calculations of pattern overlap are based on the assumption that the overlap occurs at one standard orientation. But cells will overlap with random orientation, causing a lot of variation in the binding between any two patterns. The cell surface is not a static, stable plane; time-lapse films of aggregating cells present a picture of the cell surface as a highly mobile, changing system. Another objection stems from the observation in the electron microscope that the membranes of adjacent adhering cells are mostly separated by a gap of about 100–200 A units. If the membranes seen in electron micrographs are the membranes supposedly linked by calcium bridges, then the calcium-bridge theory is in serious difficulty. The shifting of calcium atoms from a linkage within one surface to a binding link between two surfaces will not happen until the surfaces are closer than about 10 A units. If, on the other hand, the cell membrane of electron micrographs represents no more than the electron-dense central region of a membrane of much thicker dimensions, perhaps the true area of "contact" between cells is not visualized in the electron microscope and is indeed narrow enough to permit the formation of calcium bridges between opposed surfaces. Of course, if the gap is filled with cement, calcium bridges could be used to bind the cement to each cell surface on either side.

There is little doubt that calcium is important in cell adhesion. The question is how. Calcium bridges could bind cell surfaces together in some instances. Even though most unit membranes at cell surfaces are separated from those of adjacent cells by a 100-200 A gap, this is not always the case. Epithelial cells are often much more closely aligned along part of their surface, where they form so-called "tight junctions" or *zonulae occludentes* with the next cell (Fig. 4-2). In these contacts the unit membranes appear to be fused. Since the lower limit of resolution of the electron microscope is 15-20 A, cells whose membranes appear fused could be anywhere from 0-15 or 20 A apart. This is the order of separation necessary for calcium bridges, and it is tempting to suggest that calcium has a binding function in tight junctions. However, this hypothesis has not withstood the test of experiment. If calcium is depleted from the medium the adhesion of the *zonulae occludentes* is unaffected. It seems unlikely, therefore, that even these tight junctions are sites of calcium bridging. Depletion of calcium from the medium does reduce electrotonic coupling between cells, however. Thus it probably plays a role in one of the functions of the tight junction.

L. Weiss postulates that the role of calcium is to stabilize the cell surface by forming cross-linkages *within* the membrane. In his opinion, calcium depletion causes a separation of cells by its effect on the strength of the membrane. This contention is supported by the well-known observation that in the presence of calcium-complexing agents the cell surface undergoes marked activity and a characteristic blebbing.

There is also another possibility. Perhaps intercellular materials play a critical role in cell adhesion. If so, calcium bridges could be important either in binding material from opposing cells together (in the so-called "contact interspace") or in providing stabilizing cross links within the extracellular material of each cell. In either of these cases removal of calcium would cause cells to dissociate.

Curtis has suggested that calcium could aid in binding cells together in still a different way. Under physiological conditions most if not all cells have a negative surface charge. This would cause cells to repel each other:

the higher that charge, the stronger the repulsion. In order to adhere, cells must possess opposing attractive forces. These we will discuss presently. Naturally, any reduction of the repulsive forces will aid cell adhesion. Since calcium has considerable effect in reducing the surface charge of proteins and other charged particles, it is not unlikely that it has a similar effect on the cell surface. By reducing the surface charge so that the repulsive force is decreased, calcium could permit the attractive forces to pull cells closer into a firmer adhesion.

In a way, the status of the calcium problem is symptomatic of our knowledge of cell adhesions in general. It has been known for more than sixty years that this divalent cation plays an important role in binding cells together. Moreover, its importance in blood chemistry, contractility, and many other physiological processes has been established and studied intensively. Yet today, when we are trying to delineate its activity in promoting the adhesion of cells, we find ourselves only with a number of hypotheses, some of which propose utterly different activities for the ion in question. And the facts do not yet permit us to choose among these hypotheses. However, it may turn out that we do not need to choose. Calcium may well be active in several of the ways proposed, and its importance in cell adhesion could be due to a summation of its various activities.

Coordination about potassium ions

Recent studies with sodium tetraphenylboron, an agent which complexes K^+ ions, has raised the possibility that K^+ is the major cation promoting cell-to-cell adhesion in certain tissues. Rappaport has shown that liver, kidney, and brain cells of adult mice are readily and completely dissociated by treatment with tetraphenylboron. Moreover, they appear to retain their cytoplasm, including specialized protrusions such as axons, during the dissociation process. This contrasts with the action of some dissociating agents such as trypsin, which may cause considerable loss of cytoplasm. (A comparison of effects on the fine-structural level would be helpful here.) These results suggest a mechanism of aggregation by coordination of the negatively charged layers of the opposed cell surfaces about monovalent cations. K^+ appears to be the most important, but Na^+ ions may also act in this way. Na^+ ions commonly act to coordinate only when there are 6 anions symmetrically oriented in space. K^+, however, forms several types of complexes, differing both in number of anions and in their spatial arrangement. It would appear, therefore, that insofar as monovalent cations act to hold cells together, K^+ ions are the most important. Rappaport contends that the most stable cell associations would be ones in which coordination by a monovalent cation is completed by two identical sets of "fixed" anionic sites. Since cells of similar type are presumably those most likely to have identical sets of anionic sites, we have here a possible basis for selective cell adhesion and sorting out. One possible difficulty with these studies arises from the fact that the tetraphenylboron also binds calcium. Rappaport tested the dissociating properties of tetraphenylboron in the presence of calcium and found that dissociation still occurred. But it is possible that the calcium was not at a high enough molarity to countertitrate the further chelation of calcium by the reagent.

Adhesion of cells to glass

Since all cells can be made to adhere to glass and since such conditions offer the optimal opportunity for optical observations of the relation between cell and substratum, information on the mechanism of this adhesion could provide useful leads for the study of adhesions between cells. Rappaport and Taylor have both shown that cells adhere to glass within a few hours under proper conditions. This must be taken into account in evaluating studies in which observations were not made until the next day. In such cases it is probable that deadhesion is being studied rather than the initial adhesion. Rappaport discovered that the type of glass used is of critical importance, a fact overlooked by most investigators. Soft glass, for example, provides a better substratum for cell adhesion than Pyrex (of which most tissue-culture vessels have been made). It appears that attachment of a given cell to glass requires a critical number of negatively charged attachment sites per unit area of the glass surface. The cell, which is negatively charged at physiological pHs, could be attached to the negative sites by a divalent cation. Calcium is of course a prime candidate for this role, since it appears to be universally required by cells for adhesion to each other and to various substrata, including glass. After initial attachment, the cations may be displaced by protons excreted by the cells during metabolism, with consequent detachment of the cells. Continued adherence, therefore, depends on a mechanism for removing protons from binding sites. Apparently, the amount of Na^+ in the glass is a critical factor; by exchange with protons it permits a dilution of protons away from the binding sites. Significantly, Pyrex glass is low in Na^+ ions and soft glass is high.

In addition to explaining the preference of cells for soft glass, this theory also offers an explanation for one of the most serious and puzzling stumbling blocks in the many efforts to elaborate completely synthetic media for cells in culture. Most cell strains will not survive in culture without the addition of serum protein or other macromolecules to the medium. For a long time it was thought that they were needed as nutrients. This, however, appears to be wrong. The lead was given when several investigators demonstrated that large molecules facilitate adhesion and spreading on glass. Fetuin, a mixture of two proteins, was found to be particularly efficacious in this regard. Since it has now been shown that these large molecules are not necessary if a Na^+-rich glass is used, a possible mode of action of these substances is suggested. They probably promote cell survival by promoting adhesion to the cell substratum. In the presence of macromolecules, with their multivalent binding sites, the adherence of a cell to the glass surface may be the result of a protein-protein interaction. Since the protein is bound to the glass and presents an excess of binding sites to the cells, the surface properties of the glass would not be expected to be so critical.

The immediate relevance of these results to cell movement is not obvious, since there have as yet been no comparable studies on the movement of cells over glass of various compositions. We might predict that ruffled membranes would form more readily on soft glass than on Pyrex and more readily on Pyrex covered by a film of large molecules than on Pyrex not so treated.

It must be reemphasized that study of the adhesion of cells to glass does not necessarily explain their adhesion to other substrata, including other cells (see p. 72). It does, however, provide profitable leads. The suggestion that calcium ions promote adhesion to glass by forming bridges between negative sites on the surfaces of the cells and the glass is an idea with which we are already familiar. The postulated role of large molecules in adhesion is also of interest, since extracellular materials may be important in mediating cell adhesions.

Physical theories

Most if not all cells are electronegative at their surfaces, due no doubt to the presence of charged groups. Since charged groups create an electrostatic force that tends to repel charges of like sign, two negatively charged cells tend to repel each other. The question is whether the repulsion is sufficient to have an effect on cell adhesion. The most complete study of this is by Dan on the electrostatic force of sea urchin eggs. He found a close correlation between surface potential and detachment of eggs from glass under a variety of ionic conditions. Adhesiveness decreased with increasing negative surface potential. Ambrose and Easty have also found that two types of mouse tissue cells that appeared less adhesive possessed increased negative surface potential, and Ambrose found in addition that tumor cells have almost twice the negative charge of the homologous normal cells from which they are derived. This correlates with their decreased adhesion and could be a factor leading to the breaking away of cells in metastasis. Calculations of the repulsive force for various separations and surface potentials indicate that particles larger than 2000 A in diameter are unlikely to have sufficient energy from Brownian motion to approach a similar surface closer than about 60 A, if their surface potentials are -20 mV in physiological saline. Larger particles would experience more repulsion. The repulsive force forms a potential energy barrier that tends to prevent surfaces from moving together. It hardly seems coincidental that measured surface potentials of various cells vary between about -8 to -38 mV.

If surfaces have charges which are *unlike* in sign or even unequal potentials of the same sign, they will tend to attract each other. One could therefore imagine a mosaic of positive and negative charges on the cell surface, each attracting an area of opposite sign on another cell. This is considered to be unlikely on theoretical grounds, since a general electric field with differing charges would smooth out a short distance from the surface. Pethica has calculated that attractive forces between surfaces of like sign but unequal charge are unlikely to exceed 15 A in range.

Because most cells will tend to repel each other, it is necessary to postulate the existence of attracting forces in order to account for adhesion. In a search for appropriate candidates Curtis assumed that the unit membrane at the cell surface revealed in electron micrographs is the membrane that possesses the negative charges responsible for repulsion. This assumption limits one to attractive forces that oppose electrostatic forces sufficiently to establish a balance at distances of less than 40 A and at 75–200 A. Van der Waals-London forces could be an appropriate candidate. These are weak forces arising between neutral atoms. Attraction is caused by the polarization of the charge distribution of one atom, so that it is opposite to the fluctuating charge distribution in another atom, leading to a weak electrostatic attraction. Surfaces of similar constitution attract each other with their van der Waals-London forces and repel one another if they differ considerably. Electrostatic forces decline exponentially with the distance from the surface. But the van der Waals-London forces decline inversely with the square of the distance between two surfaces. As a consequence, the electrostatic forces of repulsion will approach zero and the van der Waals-London attractive forces, though weak, will remain and predominate at increasing distance from the surface. Since these forces act over considerable distances (up to several hundred angströms), they could explain the mutual attraction of cells, their adhesion at a constant separation, and the persistence of a force resisting separation of the cells. We may note in passing that this theory provides an explanation for the adhesion of cells with a considerable separation between them, without the necessity for any cementing extracellular material.

The most compelling argument in favor of van der Waals-London forces comes from a study of their interaction with the electrostatic force of repulsion in colloidal systems. It has been shown that the two forces reach a balance between two surfaces at two distances: 5–20 A and 100–200 A. Between these two regions of attraction there is a region of repulsion. The two points of balance agree strikingly with observed distances generally found between the unit membranes of adhesive cells. Ambrose and Jones have found that even fibroblasts adhering to glass are separated from their substratum by a distance. Significantly, this distance appears to approximate 100 A. Curtis showed this by examining the points of adhesion of ruffled membranes to glass with the interference reflection microscope. Finally, phospholipids, such as lecithin and cephalin, which are probably present in all cell membranes, form a stable suspension with lipid leaflets coming to be 85 A from each other when mixed with certain aqueous electrolytes. These agreements seem too close to be coincidental.

Before we become overly impressed, however, it must be recalled that there may be a third kind of cell-to-cell junction, the so-called "close junction." In these junctions the opposed plasma membranes are 20–90 A apart, the postulated region of repulsion. Yet cells appear to be firmly bound together in such regions. There is of course little or no experimental evidence for the operation of van der Waals-London forces in living systems. Knowledge of the properties that has been of use in the analysis of cellular adhesion comes almost entirely from model colloidal systems. Because of the absence of information of the operation of van der Waals-London forces in living systems the theory is at present relatively safe from criticism. Pethica (1961) has leveled several objections on physicochemical grounds; but since the assumptions which lie at the basis of his calculations are likewise based on incomplete information, Curtis finds ready rebuttal. Part of the objection is founded on Robertson's observation that hypertonic solutions of sucrose will close the separation between cell surfaces. Since sucrose has no effect on ionic processes it cannot be argued that it acts by decreasing the electrostatic forces of repulsion. One may counter by pointing out that the action of sucrose may be no more than an osmotic one, withdrawing water from between the cells and thus drawing them together despite the forces of repulsion.

It has also been objected that the gap may be filled with intercellular cementing material that does not show up in electron micrographs because it has low electron density. This is a serious possibility, in view of the low electron density of mucopolysaccharides, which are likely candidates for intercellular material, and the recent demonstrations of stainable materials in the gap (see p. 93). Curtis counters this argument by pointing out that hemoglobin will diffuse along the gap in mouse kidney tissue.

One final question may be raised. If cell surfaces differ considerably in constitution, their van der Waals-London forces theoretically would cause them to repel one another. But experiences with mixed aggregates (Ch. 7) teaches that all cells readily adhere to each other from the very beginning, regardless of type. Whatever may be the mechanism of sorting out, it is not due to one cell type repelling another. This could be used as an argument against the operation of van der Waals-London forces in cell adhesions. But the argument could also be turned in the other direction. The fact that cells of different types do not repel each other could be due to the fact that they do not "differ considerably in constitution." Since in any case we know vanishingly little about the constitution of the cell surfaces, the argument remains up in the air.

Bangham and Pethica go on to point out that the repulsive energy due to electrostatic charge is directly proportional to the radius of the particle. For example, if a radius of 0.1 μ rather than 10 μ was assigned to the particles, the repulsive energy would be considerably reduced. This would permit small particles, with a radius of the order 1000 A, to penetrate the potential energy barrier and come into close contact when driven by Brownian movement. It seems probable that this would also allow cell processes with a radius of curvature of about 1000 A to penetrate close to the surface of the opposed cell. Pethica suggests that these close contacts (of the order of 5–10 A), made initially by cell protrusions, would spread, so that extensive parts of the cell surface would "zip up" and finally come into close adhesion. They point to the end attachment of a phage to a bacterium and to the end-to-end agglutination of erythrocytes as examples of the first contact being made at a projection. Lesseps has recently attempted to test this hypothesis by observing in the electron microscope the manner in which heart and retinal pigment cells come together during the first hours after dissociation. Like other workers, he assumes that the unit membrane represents the cell surface. The surfaces of all dissociated cells are thrown into numerous undulations and villi, which closely resemble the protrusions which characterize any free cell surface (Fig. 6-1). In the light of Pethica's argument, it is of interest that their radius of curvature is approximately 1000 A. During the first few hours of aggregation, moreover, cells are seen to make initial "contacts" at the crests of these undulations and at the tips of microvilli. The fact that the initial approaches are made by processes whose dimensions meet the requirements of calculations based on model systems is important and points up the value of such calculations.

The critical question is whether these initial contacts are 5–10 A from the other cell membrane. The closest adhesions were observed to be at 120 A, more than 10 times the predicted distance. Moreover, although it is claimed that these small approaches spread and bring more of the cell surfaces into close adhesion, no micrographs are shown of this. And

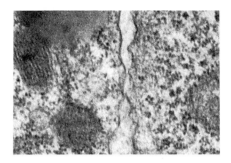

Fig. 6-1 Chick heart and pigmented retinal cells, cultured together for 15 min. The pigmented retinal cell on the left can be identified by an early-stage melanin granule (and many fully developed melanin granules not shown here); the heart cell, on the right, was identified by a glycogen body not shown in this picture. The closest points of contact are at two pairs of undulations on the adjacent cells. ×38,800. (After Lesseps. 1963. J. Exptl. Zool. **153**:171.)

even if close adhesions are later observed to occur as a result of the penetration of the repulsion barrier by fine surface protrusions, the spreading of these to form close adhesions of the whole cell surface must be relatively uncommon; the great majority of adhering cells are separated by the 100–200-A gap. In the light of these results with aggregating chick cells it should be noted that aggregating cells of sea urchin mesenchyme blastulae do not behave in the same way. These cells show no such microprotrusions, even in material fixed as early as 10 minutes after the beginning of aggregation. Cells at this time are already flattened against each other at a distance of approximately 150 A. This observation emphasizes the need for more such studies before we are in a position to generalize. It may be said, therefore, that whereas Pethica's postulations may account for the adhesions of certain cells, they do not account for all.

There are other physical forces that may also play a role in cell adhesions. At present these exist only as possibilities derived from the physical chemistry of nonliving systems. No attempts at biological analysis are yet in evidence.

It may be asked where specialized intercellular junctions, such as desmosomes and septate desmosomes, fit into this physicochemical world. Essentially nothing is known, and there has been very little speculation. Because of the dimensions of the gap in these junctions, van der Waals-London forces may be at work. It is also possible that these junctions are not responsible for initial cell contacts but form later, to solidify the adhesion. In this regard it is of interest that early chick and teleost embryos lack desmosomes but form them later.

Isolation of cell membranes

One of the main difficulties with theorizing on the manner in which cell membranes come together is our lack of information on the chemical constitution of the membrane. Considering its importance for the cell, not to speak of for the cell biologist, it is genuinely surprising how little reliable information is available on the composition of the cell membrane. There is good evidence that it contains lipids, such as lecithin, cholesterol and sphingomyelin, and protein, but little is really known about how they are organized. It seems certain also that mucopolysaccharides compose an important part of most membranes and especially of intercellular materials. Attempts to analyze the composition of the plasma membrane ultimately require the development of suitable techniques for stripping it of its underlying cytoplasm and other tests for showing whether or not it is indeed the membrane that has been isolated. Erythrocyte "ghosts" have been available for study for many years because their surface structure is strong enough to remain intact during isolation. But because of the special nature of the erythrocyte its membrane has less interest for us than corresponding ghosts of the cells of cohesive tissues. Besides, erythrocyte membranes have been demonstrated to contain nonplasma-membrane material. Unfortunately, the surface membranes of most other kinds of cells seem not stable enough to remain intact while the nucleus and cytoplasm are being flushed out. The main criteria for identifying fractions consisting of the cell surface have been lipoprotein composition and appearance in the electron microscope. A method just developed that appears to work effectively for the L-cell (a well-known line of tissue-culture cells) is first to strengthen or "tan" the cell periphery with fluorescein mercuric acetate or various metallic ions used in hypotonic solution. This causes the cell membrane to rise off its underlying cytoplasm. The cells are then ruptured in a homogenizer and the membranes fractioned by differential centrifugation. This method yields clean membranes with the typical trilaminar structure seen under the electron microscope in sufficient quantities for detailed analysis. Some membranes isolated with this technique still have desmosomes.

Another method is to homogenize cells in a medium containing dilute citric acid or increased $CaCl_2$. When this method is applied to minced pieces of rat liver, several membrane fractions are obtained which can then be purified by low-speed differential centrifugation and flotation in a discontinuous sucrose gradient. This approach has the advantage of yielding membranes from tissue cells in normal contact relations with each other. As judged primarily by their appearance in the electron microscope, the plasma membranes obtained by this method are remarkably clean and altered very little morphologically. They show the typical trilaminar structure, tight junctions, close junctions, and desmosomes. Moreover, when subjected to negative staining, they even show the hexagonal subunit pattern typical of plasma membranes at tight junctions.

The desired chemical information has now begun to be extracted and soon should be considerable. Much further refinement of the analytical techniques will be necessary, however, before membrane isolations can aid in elucidating the chemistry and structure of the membranes of cells during morphogenesis. At present the methods are amenable only to a limited range of cells, and the quantities of cells of one type that can be obtained from early embryos for membrane isolations are far too small to permit the desired chemical analyses.

In the meantime, one compound that has been found to occur in the external surface of many, if not all, cells is sialic acid (more correctly known as N-acetyl-neuraminic acid). It is a 9-carbon sugar with a strongly acidic carboxyl group. Sialic acid is liberated by neuraminidase, and the use of this enzyme on whole cells has indicated that sialic acid is highly important in cell behavior. It has been established, for example, that treatment of red blood cells and a variety of human cells in culture with neuraminidase reduces surface charge. That sialic acid is not only involved in influencing surface charge but may have other effects on cell behavior is suggested by a pretty study by Forrester. A clonal line of normal hamster cells, which is stable in cell culture and shows contact inhibition, was infected with polyoma virus and transformed into a stable line of cancer cells, which is not contact inhibiting. When the electrophoretic mobility of clonal lines of these normal and tumor cells was measured (Fig. 6-2), the polyoma transformed cells were found to be decidedly more electronegative. Treatment with neuraminidase depressed the surface charge of both cell lines, as expected, but reduced that of the tumor cells more, so that its charge became the same as that of the normal cells. These results show that the

Fig. 6-2   After J. A. Forrester. 1965. In E. J. Ambrose [ed.] Cell Electrophoresis. Table IV, p. 122.

Clone	Mobility before neuraminidase treatment ($\mu/sec/v/cm$)	Mobility after neuraminidase treatment ($\mu/sec/v/cm$)
Normal cells	-1.00 ± 0.05 -0.97 ± 0.06	-0.67 ± 0.05 -0.64 ± 0.05
Transformed cells	-1.30 ± 0.06 -1.27 ± 0.05	-0.65 ± 0.06 -0.65 ± 0.05

enhanced electrophoretic mobility of the transformed cells is due to increased concentration of sialic acid residues in the cell surface; and since surface charge has been shown to be related to cell contact behavior (see p. 83), it seems likely that sialic acid is involved in some way in the different contact behavior of these two kinds of cells.

Intercellular materials

Materials have been observed between cells for a long time. It was, therefore, natural to believe that they were responsible for cell adhesion. This intercellular material varies enormously in quantity and quality from one tissue to another. Some cells produce so much that the tissue or organ comes to be composed mostly of the intercellular material. So it is with cartilage, bone, and the fibrous connective tissues. The fact that collagen, the extracellular product of fibrocytes, makes up about 30% of the body protein is ample illustration of the importance of intercellular material in the architecture and economy of the organism. In cartilage, bone, and the fibrous connective tissues the cells are so widely separated by their own products that it has been thought that these materials must surely be the main factor keeping cells in position and thus preserving the integrity of the tissue. In other tissues, where cells are so tightly packed that no separation can be seen with the light microscope, the situation is not so obvious. Nevertheless, so ingrained was the idea that if cells stick together you must have something to stick them with, that a cementing substance was simply assumed to be there. It was called "cell cement."

The belief in cell cement was based in part on a substantial body of careful research. It has been known for a long time that the reason blastomeres of echinoderm eggs may be dissociated if placed in calcium-free seawater is because in the absence of calcium a transparent, extracellular surface layer of the egg is dissolved away. This layer came to be known as the hyaline plasma layer (Fig. 6-3) and has been shown in many studies to be the main factor binding blastomeres together to form a normally organized spherical cleaving mass. It is a product of the egg itself and is formed of material exuded from the egg surface soon after fertilization. An outer layer similar to this may be present and play a similar role in many eggs.

In the era of the electron microscope a simple belief in cell cement no longer suffices to account for the adhesion of cells. Most cells are so close together that we become involved in discussions as to whether intermolecular bonding or long-range physical forces are involved. There has nevertheless been a revival of the notion that cells may be bound together by intercellular materials. This has been caused primarily by three kinds of

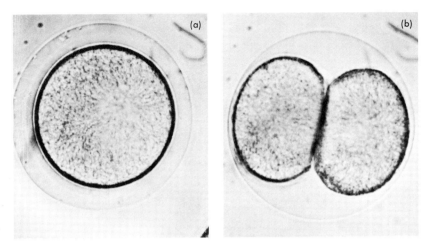

Fig. 6-3 (a) Fertilized egg of the sea urchin, *Paracentrotus lividus*, showing the clear hyaline plasma layer on the surface of the egg. The membrane some distance off the egg surface is the vitelline membrane. ×70. (b) A two cell stage of the Paracentrotus egg, showing the hyaline plasma layer uniting the two blastomeres. ×70. (Trinkaus. Unpublished.)

observations: 1) the dissociation of cells by the action of enzymes, 2) the discovery of the famous 100–200-A gap separating the electron-dense plasma membranes of many tissue cells, and 3) the production of viscous exudates by cells in culture under various conditions.

The efficient dissociation of cells by enzymes such as trypsin, pancreatin, papain, elastase, collagenase, and mucase has been often interpreted as due to the rupture of chemical bonds which hold cells together. Since the cells survive and with gentle treatment appear to be largely intact, it was quite naturally thought that the enzymes could act by digesting cementing material between cells. This impression has been fortified by the frequent appearance of a slimy exudate in the medium during the dissociation procedure. But this material has been shown in one case to consist largely of DNA. The validity of this hypothesis also depends on the mode of action of the enzymes during the dissociation process. In spite of the fact that trypsinization is one of the most widely used techniques in cellular biology today, it is not all certain that it or other enzymes dissociate cells by disrupting chemical bonds. The dissociation of cells by enzymes therefore cannot be taken as automatic support for the idea that cells are held together by intercellular materials.

The 100–200-A gap has sometimes been seen to show slight staining after fixation with osmium tetroxide, suggesting the presence of a "gap substance." Even when such material appears, however, one is not justified in concluding that it is normal extracellular material. First of all, it may simply be the less dense part of the cell membrane (refer to p. 59). Secondly, we have no assurance that the processes of fixation do not leach material from inside the cells and deposit it in the gap. The lack of penetration of ferritin particles between cells after fixation is suggestive in this regard. Finally, there is the sucrose experiment of Robertson, already described. If the gap can be closed by application of an osmotically active compound like sucrose, what assurance do we have that the gap is not largely filled with a watery solution of electrolytes? As a matter of fact, if high-molecular-weight material between cells reaches an appreciable concentration, is it not likely that appreciable water uptake would occur, causing the gap to widen? In addition, if there is cementing material between cells, would we not expect it to vary in thickness? It varies enormously in thickness in tissues, where its presence is readily demonstrated with differential staining under the light microscope. This thought suggests that the striking constancy of the dimensions of the gap may itself be an argument against it being a repository for intercellular cementing material. Where the gap between cells is greater than 200 A, and there are certainly many examples of this (Figs. 14-10 and 14-12), it is quite possible that cementing material is necessary. Even *if* long-range van der Waals-London forces are important at 100–200 A, they would be expected to weaken drastically at greater distances. And no other long-range forces are known which would be sufficiently strong to attract cells over a distance greater than about 200 A, given the electrostatic properties of the cell surface. Incidentally, even if intercellular material is present in a gap of 100–200 A, it may have been secreted later, after adhesion had occurred. By chemical bonding it would then provide an added means of holding the cells together.

A type of cell contact that may involve intercellular materials is the desmosome (Fig. 4-2). This structure, often found at cell junctions in epithelial tissues, is apparently in part responsible for the tight cohesion of epithelial cells. In osmium and $KMnO_4$-fixed material, the gap substance between two opposing desmosomal plaques is often considerably more electron dense than elsewhere along the gap. The plaques are usually about 200 A apart. However, the true significance of this observation is rendered questionable by the fact that both hemoglobin and ferritin diffuse into and through the central region between two desmosomes.

It should be clear from these considerations that evidence from osmium- and $KMnO_4$-fixed material viewed in the electron microscope is not enough. We need to know of what the gap is composed. Perhaps techniques now being developed for doing cytochemistry at the fine-structural level will provide some answers. The first results with lanthanum and periodic acid-silver methenamine (PASM) staining are encouraging. The lanthanum-stained material appears to vary quite a lot in thickness, but the PASM-stained material seems to have a rather uniform width, averaging 140 A. In both cases stained material is not present in tight junctions. The argument is thus strong that the materials are extracellular and, in the case of the PASM material, fill the 100–200-A space. It is not yet established, however, whether these materials were there when the cells first adhered or were deposited later, whether they are necessary for cell adhesion, and finally whether they are not in part artifacts of the fixation process, composed in part of materials leached from the cells. It should be recalled that, although ferritin particles penetrate the 100–200-A gaps of living tissues, they do not penetrate the gap after fixation.

The groundswell of current interest in the possible activity of intercellular material in cell-to-cell adhesion was given its momentum by certain fundamental studies of other cell activities. The contact guidance theory of P. Weiss is the first of these. It was based, as we know, on the behavior of cells in culture moving under the guidance of an artificial substratum. But why is it, he asked, that spindle cells in culture will orient on a solid glass or plastic fiber or groove in the direction of the fiber of a groove axis? There is no reason to believe that a glass or plastic surface would offer orientations to which a cell could respond. Close observation suggested a possible answer.

Cells were observed to produce macromolecular exudates (called "ground mats" by Weiss) which spread along the interface between the cell and its substratum. On a cylindrical fiber the exudates must advance mainly in the forward direction along the axis of the fiber. According to Weiss, this will stretch the meshes of the micellar network behind in a lengthwise orientation. The leading edge of the cell may then trace these tracks, and the rest of the cell will follow, moving lengthwise along the fiber. Curtis points out, however, that the surface tension effects controlling the microexudate would be contrary on ridges and grooves and should therefore have different orienting effects, if the orientation of the microexudate does indeed orient cells. It is conceivable that cellular exudates are also oriented in the organism and provide the cells with an oriented substratum to guide their movements. Rosenberg has recently found that cells in culture do indeed release macromolecular materials that coat the substratum with monomolecular films. Although studies demonstrating the postulated orientations in the ground mat and directed cellular responses to them still remain to be done, the idea is a provocative one and has gained acceptance as a good working hypothesis. As such it has kindled much of the contemporary interest in the importance of cellular exudates for cell contact.

In similar fashion, the work of Grobstein on the mechanism of embryonic induction has contributed to current interest in the activities of large molecular materials exuded from cells. By means of millipore filters of varying thickness and pore size he was able to show that induction of kidney tubules by mesenchyme or dorsal spinal-cord inductor will take place even when cell surfaces are separated by distances of the order of 50 μ. Even though extensions of the cell surface penetrate the larger pores, they fall far short of the surfaces of the opposing cells. Extracellular material, apparently composed in part of mucopolysaccharide, is found in this gap. The critical material appears to be macromolecular, since its passage is blocked by cellophane and is impeded by a pore size of 100 A. These facts led Grobstein to postulate that a kind of "intercellular matrix" transmits morphogenetic information during induction. This concept of the matrix was a useful idea, since it was part of an attempt to look quantitatively at the induction process and to formulate a testable working hypothesis. As such, both the work and the idea it generated represented a new look at an old problem and caused all of us to contemplate with renewed interest the possibility that there exists external to and in between cells a material that possesses both stability and specificity and that is the first line of activity in the controlled interactions between cells during differentiation.

With this background, it was natural that the slimy material that often appears in the medium when cells are dissociated with trypsin and EDTA should excite interest in its possible intercellular origin. Moscona in particular was struck by this thought and has performed experiments designed to determine whether this material promotes cell aggregation. If dissociated cells are rotated at 38°, they cohere rapidly and form large aggregates. At 25° they form only minute aggregates. If cells are incubated at 38° for two hours but prevented by rapid rotation from aggregating, they liberate a material containing mucoprotein into the medium. If freshly dissociated cells are rotated at 25° in a medium conditioned by this material, they aggregate more completely than in fresh medium. Moreover, according to Moscona, this material shows a certain amount of tissue specificity. Material secreted into the medium by neural retinal cells promotes aggregation of neural retinal cells, and to a lesser degree cells of the skin and the liver, but has no effect on aggregation of kidney or limb mesenchyme cells. From results such as these, it has been concluded that selective cell association may involve specific intercellular materials synthesized by the cells themselves. Of course, to show specificity rigorously, converse experiments with substances produced by other tissues are needed.

That the material in question arises from the neural retina cells is indisputable; the question is whether it is exuded from inside the cells and is therefore intracellular in origin, or whether, as contended, it is removed from the cell surface and represents intercellular adhesive material. The presence of mucoprotein in the conditioned medium supports the latter idea, since such protein is often associated with the cell surface. This is not sufficient evidence, however, because it has been shown that under certain conditions of dissociation cells may diminish in volume and lose much material. Incorporated P^{32}, for example, appears in the medium after trypsinization, suggesting that nucleotides or polynucleotides are released from the cells. In a test of this hypothesis, Steinberg found that the slimy material can be removed from the medium by crystalline deoxyribonuclease. This suggests that the material is rich in DNA, hardly a molecule ordinarily associated with the cell surface. Moreover, more slime is produced if the tissue is cut into finer pieces prior to dissociation; the more cell injury, the more slime. This observation is also consistent with an intracellular origin of the material. Finally, when the slime is digested away either with deoxyribonuclease or with crude pancreatin (which contains deoxyribonuclease), the ability of the cells to aggregate is not impaired.

But the story does not end there. Moscona and his associates have pursued the problem further and Lilien in particular has contributed convincing evidence that the supernatant from the culture medium of monolayer cultures of neural retina cells contains a deoxyribonuclease resistant factor(s) that enhances the aggregation of freshly dissociated retinal cells. That this material in the supernatant is tissue-specific is supported by four lines of evidence: 1) cells from tissues other than neural retina do not respond to it, 2) supernatants prepared from cultures of other types of cells do not promote the aggregation of retina cells, 3) the aggregating activity of retina supernatent may be removed by neural retina cells but not by liver cells, and 4) antiserum prepared against retina supernatant modified the normal aggregation of retinal cells but not that of liver cells. This last bit of evidence suggests that the active material is indeed at the cell surface. Whether it is "intercellular material" or not is still a moot question. Supernatants from other tissues thus far show no tissue-specific activity.

Observations by two workers in P. Weiss' laboratory are pertinent here. Rosenberg found that cells exude a mucoprotein material on a chemically clean surface when cultured in protein-free media. If the substratum has an absorbed layer of protein, however, no exudate appears. Adhesion occurs anyway, with or without a microexudate. The necessity of a layer of protein between cells and Pyrex glass for what appears to be normal adhesion is substantiated by the observations of Taylor. He found that both live and dead (formol-fixed) cells will attach rapidly and firmly to glass in a protein-free medium. This adhesion is not calcium dependent and is not inhibited by trypsin. If protein is added to the medium or if the glass substratum is coated with protein, only live cells adhere (but more slowly). This adhesion is like cell-to-cell adhesion in that it is disrupted by both EDTA and by trypsin. Both of these observations are consistent with the conclusion that a layer of protein is necessary to obtain adhesions to glass which approximate cell-to-cell adhesions. But the protein need not be produced by the cells concerned. The argument therefore favors the necessity of a protein layer but is against its type specificity.

Let us now turn briefly to the sponges, where some interesting studies have been made. We include them here because the authors believe they are studying intercellular material. Humphreys and Moscona have found that mechanically dissociated sponge cells reaggregate rapidly at both room temperature (24°C) and at 5°C. If, however, the cells are chemically dissociated by means of calcium- and magnesium-free seawater, cells will not reaggregate at 5°C, even when calcium and magnesium are restored to the medium. It would appear that chemical dissociation removes something from the cells that is essential for their aggregation and cannot be restituted at 5°C. In the hope that this hypothetical factor is released into the supernatant in an active state, Humphreys collected supernatant and added it, plus calcium and magnesium, to cells at 5°C. They formed compact clusters. Then he moved on to perform a crucial experiment. One of the well-known features of sponge reaggregation has been the high degree of species specificity displayed. In a mixture of dissociated cells of *Microciona* and *Haliclona*, for example, cells of the two species sort out and aggregate only with those of the same species. If the factor extracted from sponge cells is closely associated with aggregation per se, we would expect it to show similar species specificity. Sure enough, supernatant from *Microciona* was found to be active in promoting aggregation of *Microciona* cells but not cells of *Haliclona*, and vice versa. (This was a stroke of good luck, for subsequent investigation of extracts from other sponges has revealed little of this species specificity). Since cell adhesion is a cell surface phenomenon and since the extracted factor displays the same species specificity as cell adhesion, it would appear likely that the aggregative factor in the supernatant is somehow associated with the cell surface. The authors incline toward identifying it with intercellular material. This seems premature. At this stage of the investigations it could as well be membrane material or intracellular material as intercellular material. As a matter of fact, it is even possible that its effect is simply to reverse inhibition of certain synthetic reactions, which then permits the cells to aggregate. Unlikely though this may appear, neither it nor other possibilities will get proper treatment until we know something of the nature of the active principle(s); its activity in promoting aggregation; and the changes, both structurally and functionally, in sponge cells after its removal. Regardless of the outcome, the results are of interest, for this is the first time that a material that clearly promotes specific cell aggregation has been collected in a cell-free state.

By way of general conclusion, it is undeniable that intercellular material exists, but, except for the massive tissue-forming materials such as collagen and cartilage, we know little of its composition. Nor has there been any clear demonstration of its role. The problem is an excessively difficult one because of the tendency of cells to exude large molecules in culture and during fixation. In future research the essential question will not be whether cells produce exudates that influence the contact behavior of other cells. Many substances influence cell adhesion and cell locomotion in diverse and obscure ways. The questions rather will be: 1) the chemical nature of these exudates, 2) whether these exudates have specificity or characteristic quantitative differences that correspond to the cell type or origin, 3) whether the exudates can maintain this stability external to the cells which produced them, and 4) whether the material is in fact extracted from the surface region of the producer cell or is leached from the interior.

Selective deadhesion

A possible mechanism of selective cell adhesion that has received little attention deserves brief consideration. L. Weiss has also been impressed with the fact that initially in a mixed aggregate and generally in culture all cells adhere to each other or to an inert substratum, provided the conditions generally necessary for adhesion apply. In physiological environments conditions are apparently just right for initial adhesive contacts to take place among cells. How then can we explain sorting out in a mixed aggregate of cells? Cells which initially adhere to those of another type must obviously break away to make new adhesions with cells of their own type. Since all cells possess at least a limited amount of mobility, one may assume that if the adhesion does not provide enough mechanical restraint to movement, a cell will disrupt the adhesion and continue its movement. The more restraint, the less chance that a cell will *deadhere* and move on. The distinctiveness of Weiss' hypothesis is the emphasis on selective deadhesion rather than on adhesion, which is assumed to be general and essentially equal.

The mechanical restraints to movement can at this time be ascribed to a number of mechanisms already discussed, but Weiss favors a thoroughly novel proposal. He suggests that the restraint lies in the mechanical properties of the cell surface or cell membrane and its resistance to rupture, rather than in the strength of adhesion between a cell and its substratum. The evidence for this idea has already been described and discussed (p. 70). It need only be added here that all the evidence is derived from the adhesion of cells to inert substrata. Cell-to-cell adhesion, which is of paramount interest and which indeed the theory is intended to explain, has not been studied. Until this is done and has provided convincing evidence that the cell surfaces of different cell types do indeed vary consistently in their natural cohesiveness, this theory must rest where it is, somewhere aside of the mainstream.

SELECTED REFERENCES

Ambrose, E. J. [ed.] 1965. Cell electrophoresis. Little, Brown and Co., Boston. 204 p.

Ambrose, E. J. and G. C. Easty. 1960. Membrane structure in relation to cellular motility. Proc. Roy. Phys. Soc. Edinburgh **28**:53-63.

Bangham, A. D. and B. A. Pethica. 1961. The adhesiveness of cells and the nature of the chemical groups at their surfaces. Proc. Roy. Soc. Edinburgh, B. **28**:43-52.

Benedetti, E. L. and P. Emmelot. 1968. Structure and function of plasma membranes isolated from liver, p. 33-120. In A. J. Dalton and F. Haguenau [ed.] The membranes, vol. 4 of Ultrastructure in biological systems. Academic Press, New York.

Crandall, M. A. and T. D. Brock. 1968. Molecular aspects of specific cell contact. Science **161**:473-475.

Curtis, A. S. G. 1962. Cell contact and adhesion. Biol. Rev. Cambridge Phil. Soc. **37**:82-129.

Davis, B. D. and L. Warren [ed.] 1967. The specificity of cell surfaces. Prentice-Hall, Inc., Englewood Cliffs, N.J. 290 p.

Grobstein, C. 1961. Cell contact in relation to embryonic induction. Exptl. Cell Res. Suppl. **8**:234-245.

Herbst, C. 1900. Über das Auseinandergehen von Furchungs- und GewebeZellen in kalkfreiem Medium. Arch. Entwicklungmech. Organ. **9**:424-463.

Humphreys, T. 1963. Chemical dissolution and in vitro reconstruction of sponge cell adhesions—I isolation and functional demonstration of the components involved. Develop. Biol. **8**:27-47.

Lesseps, R. J. 1963. Cell surface projections: their role in the aggregation of embryonic chick cells as revealed by electron microscopy. J. Exptl. Zool. **153**:171-182.

Lilien, J. E. 1968. Specific enhancement of cell aggregation in vitro. Develop. Biol. **17**:657-678.

Millinog, G. and G. Giudice. 1967. Electron microscopic study of cells dissociated from sea urchin embryos. Develop. Biol. **15**:91-101.

Moscona, A. A. 1962. Analysis of cell recombinations in experimental synthesis of tissues in vitro. J. Cellular Comp. Physiol. **60,** Suppl. **1:**65–80.

Pethica, B. A. 1961. The physical chemistry of cell adhesion. Exptl. Cell Res. Suppl. **8:**123–140.

Rappaport, C. and G. B. Howze. 1966. Dissociation of adult mouse liver by sodium tetraphenylboron, a potassium complexing agent. Proc. Soc. Exptl. Biol. Med. **121:**1010–1016.

Rappaport, C., J. P. Poole and H. P. Rappaport. 1960. Studies on properties of surfaces required for growth of mammalian cells in synthetic medium. Exptl. Cell Res. **20:**465–510.

Steinberg, M. S. 1958. On the chemical bonds between animal cells: a mechanism for type-specific association. Am. Naturalist. **92:**65–82.

Tyler, A. 1946. An autoantibody concept of cell structure growth, and differentiation. Growth. **10,** Symp. **6:**7–19.

Weiss, L. 1962. The mammalian tissue cell surface, pp. 32–54. *In* D. J. Bell and J. K. Grant [ed.] The structure and function of the membranes and surfaces of cells, Biochem. Soc. Sympos. No. 22. Cambridge Univ. Press, Cambridge.

Weiss, P. 1947. The problem of specificity in growth and development. Yale J. Biol. Med. **19:**235–278.

Weiss, P. 1958. Cell contact. Inter. Rev. Cytol. **7:**391–423.

SEVEN

Cellular Segregation and Selective Adhesion

Warren Lewis, a pioneer in cell biology, set the theme for one of the dominant efforts of contemporary research when in 1922 he wrote that the adhesive quality of tissue cells is a subject of immense importance that up until that time had been almost entirely ignored. He went on to point out that on this adhesiveness rests the integrity of tissues and organs and finally of the organism itself. As he put it: "Were the various types of cells to lose their stickiness for one another and for the supporting extracellular white fibers, reticuli, etc., our bodies would at once disintegrate and flow off into the ground in a mixed stream of ectodermal, muscle, mesenchyme, endothelial, liver, pancreatic, and many other types of cells."

Reconstitution of cellular aggregates

Actually, the questions to which we will devote our attention in this chapter were already posed implicitly near the beginning of the century in some remarkable experiments on sponges. H. V. Wilson cut adult specimens of the red Atlantic sponge, *Microciona prolifera*, into pieces and then carefully squeezed them through fine bolting silk into sea water. The result was a suspension of cell fragments, whole individual cells, and small cell clusters. This totally disorganized population slowly settled to the bottom of the dish, where the individual cells and small clusters engaged in active movements on the glass substratum. As cells contacted each other they tended to adhere, and in this manner most cells gradually became concentrated into aggregates of varying size. Some of these aggregates fused, and the large aggregates that resulted soon became spherical. Then, during the second day, they flattened on the glass substratum and coalesced with adjacent aggregates. During the next several days a remarkable process occurred. Spicules and canals formed, and even flagellated chambers lined with choanocytes. Finally, after two to three weeks, spongin was present, the canals had branched, and an osculum appeared. Reaggregates of dissociated sponge cells formed little sponges! This phenomenon was termed "reconstitution" and, aside from its intrinsic interest as a remarkable process, was considered convincing evidence of the primitive phylogenetic position of the sponges. It was therefore no surprise that other sponges, and certain hydroids as well, were soon found to possess the same capacity.

For all the wonder excited by these experiments, the mechanism remained a mystery. One of the most complete investigations of the aggregation process was by Galtsoff, who noted that small aggregates appear to be formed when the large amoeboid archaeocytes move around, collecting other cells by adhering to them. He, and others also, showed that aggregation was not due to mere stickiness of the cell surface, for when cells of two different species such as *Microciona* and *Haliclona* were mixed, they adhered only to cells of their own species. What goes on within the aggregates during the process of reconstitution is still unknown. With only histological evidence to go on, Wilson thought he saw transitional stages between different cell types and suggested that cell transformations were occurring. Others, like Huxley, assumed that the various cell types migrate within the aggregate and adhere to each other differentially, each to others of its same type. By this means the different tissue types could sort out and then become rearranged to reestablish the normal architecture of the organism. Proof of one or the other of these theories awaited separation of different cell types so that the initial composition of the aggregate could be known. However, since the various kinds of cells and tissues are quite intermingled in an adult sponge, most everyone has been discouraged from trying to separate them. Moreover, sponge cells are difficult to distinguish with certainty in the dissociated state and cannot be followed during reconstitution. Huxley succeeded in separating out pieces of gastral lining and found that these rolled up into spheres of collar cells, which during up to two weeks of culturing gave rise to no other kinds of cells. But the collar cells were not labeled and mixed with other cell types to see if they would sort out or be influenced in their differentiation by cells adjacent to them. The conclusions of both cell transformation and sorting out according to original type, therefore, have rested less on evidence than on preconceived notions of the degree of fixity of differentiated cells.

Reconstitution was considered the private province of lower invertebrates, such as sponges and hydroids, until relatively recently. Perhaps this is why such an extraordinary phenomenon excited so little investigation. Also, developmental biologists were more interested in embryos, and histologists more interested in vertebrates. Whatever the reason, with the demonstration twenty-five years ago that dissociated cells of higher organisms can also reconstitute tissue and organs, investigation of the process began in a number of laboratories. These modern investigations had their inception in Holtfreter's discovery in 1943 that cells of an amphibian gastrula will dissociate when exposed to saline solution at pH 10 and will reaggregate to form firm masses when the pH is reduced to 7+. If prospective pronephros cells are treated in this manner, they reaggregate and form more or less complete pronephric systems with nephric tubules and blood vessels. When some years after Moscona obtained characteristic tissue reconstitution in aggregates formed from mesonephric cells or chondrogenic cells of 4-day chick embryos that had been dissociated by trypsin, it appeared that reconstitution by dissociated cells is a phenomenon of general biological significance. As such, it was yearning for explanation. For this purpose vertebrate embryos were more hopeful material than sponges, because of the ease of separating organs and even certain tissues.

Rather than proceed directly to the analysis of the behavior of dissociated cells, it will be instructive at this juncture to pause and examine certain experiments on intact tissues that have provided insights helpful to the analysis. In 1939 Holtfreter published a paper now recognized to be one of the fundamental studies of early development. He separated the germ layers of amphibian gastrulae and then reunited them in culture in various combinations. Two pieces of pure endoderm fuse to form a smooth sphere. Presumptive epidermis and endoderm at first fuse closely, but later they move apart until only a thin bridge connects them. In Holtfreter's terminology endoderm has "positive affinity" for endoderm but "negative affinity" for ectoderm. Since their behavior in vitro corresponds to their behavior in vivo, a critical experiment would be to mix pieces of endoderm and ectoderm with mesoderm. The mesoderm adheres to both endoderm and ectoderm (positive affinity) and thus unites two nonadherent layers. In Holtfreter's experiments, mesoderm was always on the inside and ectoderm always on the surface. Endoderm was sometimes also on the surface, held to the ectoderm by its adhesion to mesoderm, and sometimes in the interior, where it formed a tubular gut (Fig. 7-1). The latter condition was strikingly normal in its organization; the gut was embedded in a mass of mesenchyme, which in turn was covered by a surface layer of ectoderm. These experiments are significant for two reasons. First, they

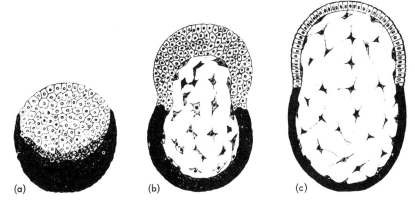

Fig. 7-1 Combination of amphibian endoderm and ectoderm with a layer of prospective connective tissue interposed. (a) Close union of the cells, (b) protrusion of the endoderm, (c) spreading of the ectoderm and endoderm to form a common epithelial wall around an internal cavity filled with mesenchyme. (Holtfreter. 1939. Arch. Exptl. Zellforsch. 23:169.)

proved that it takes more than mere stickiness to make germ layers adhere. It is rather a highly selective affair, in which the layers adhere to each other differentially. In addition, from the fact that mesoderm is necessary to keep ectoderm and endoderm together, it appears that these affinities, or selective adhesions, are important in normal morphogenesis.

More recently Chiakulas, in Weiss' laboratory, showed that the same selectivity characterizes fully differentiated tissues. With elegant simplicity he removed an area of skin from the flank of a salamander larva and grafted various epithelia in its place. A graft of epidermis spreads over the denuded surface, fuses readily with the host epidermis by a cellular intermixing at the juncture, and ceases migration. Tissues normally contiguous with epidermis, such as cornea and oral epithelium, also fuse in this manner, but not noncontiguous tissues (esophagus, small intestine, bladder, etc.). They also adhere at the point of contact, but the adhesion is different and involves spreading of one epithelium over the other. By ringing a rod of cartilage with two tissues and thrusting the whole assemblage into the gelatinous interior of the dorsal fin, he again found fusion and cellular intermingling only between pieces of the same tissue or normally contiguous tissue, e.g., esophagus and esophagus, esophagus and lung, or esophagus and oral epithelium, but not between noncontiguous tissues such as lung and oral epithelium. Obviously, both early and advanced tissues have capacities for selective adhesion. By this means they can distinguish not only their own kind but have the additional talent of recognizing some of the other tissues with which they must normally associate. They possess both isoaffinity and heteroaffinity.

It is important to emphasize the importance of both kinds of adhesions.

The adhesion of like to like is necessary to establish and maintain the integrity of individual tissues and is probably the reason why like tissues make firm contacts after grafting operations, even though normal topography is drastically disrupted. Some years ago, for example, Hooker found that after rotation of a piece of the embryonic spinal cord up to 135°, the central canal and fasciculi exhibit just the right amount of torsion necessary to reestablish continuity of each of the areas. In his studies of neuronal connections within the central nervous system, Sperry has found that regenerating nerve fibers show an extraordinary degree of specificity. The terminal connections of growing nerve fibers are laid down in a highly selective fashion and are apparently governed by the intrinsic specificity of the advancing fiber tip plus that of the various cellular elements it encounters along its route of outgrowth. For example, when medial and lateral bundles of the optic nerves are cut and crossed, regenerating fibers do not enter the optic tectum indiscriminately or along lines of apparent least resistance (Fig. 7-2). They promptly recross against the imposed mechanical biases to reenter their own proper channels.

The adhesion of unlike tissues lies at the basis of organ architecture and must play an important role in organogenesis. The adhesive contacts of pseudopodal extensions of mesenchyme cells to the blastocoel wall in sea urchin gastrulae (p. 33) is an example of adhesion of unlike cells which has unquestioned importance in morphogenesis. An obvious case of selective affinity between unlike cells that has been recognized for a long time is the union of the sperm and the egg.

Now let us return to the problem of organ reconstitution by dissociated cells. In an effort to see if cells from diverse sources within the organism would influence each other's differentiation when in advanced stages of differentiation, Peggy Groves and I dissociated cells of the chick mesonephros and wing bud by the trypsin method of Moscona and then recombined them by centrifugation. In spite of the fact that cells of one type were in intimate contact with those of another type within the mixed aggregate, no new modes of differentiation appeared. Both mesophrenic and limb-bud tissues were reestablished, often in particular arrangements, with keratinizing epidermis on the periphery, cartilage in the center, and mesonephric tubules between the two (Fig. 7-3). Since no way had been found to mark these cells, all one could conclude was that here, too, in a mixture of different cell types, reconstitution occurred. The particular origin of each cell in the differentiated aggregate remained unknown.

At the same time, Holtfreter was extending his studies with cells of amphibian embryos. Here it was possible to distinguish the germ layers from each other to some degree. Ectoderm cells, for example, are pigmented and small, whereas endoderm cells are unpigmented and large. With his student Townes, Holtfreter took advantage of these differences in a brilliant

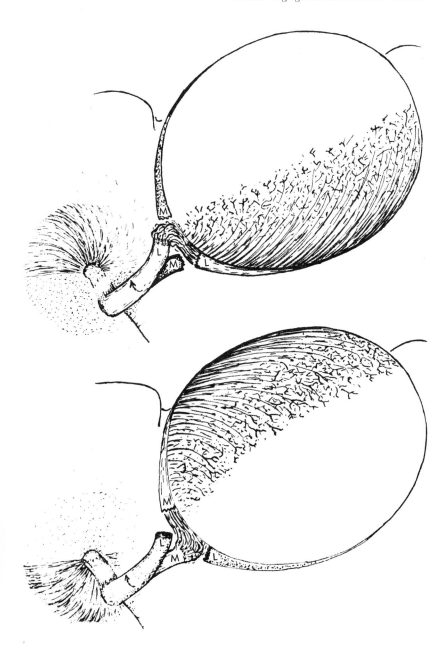

Fig. 7-2 Experiment in which the medial and lateral bundles of the optic nerves are cut and crossed. Regenerating fibers do not enter the tectum indiscriminately or along lines of least resistance but recross to enter their proper channels. (Arora and Sperry. 1962. Amer. Zool. **2**:389.)

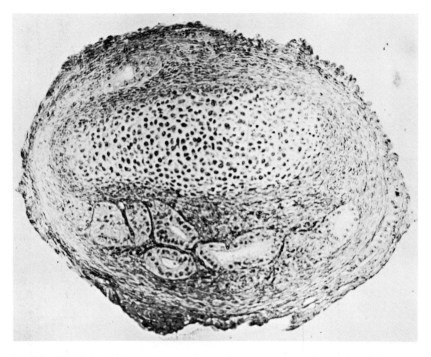

Fig. 7-3 Explant derived from a mixed aggregate of approximately equal quantities of dissociated cells from 5-day chick mesonephros and 5-day wing bud. Note regional differentiation of cartilage and mesonephric tubules. Ten days in culture. Stained with Mallory's triple stain. ×160. (Trinkaus and Groves. 1955. Proc. Natl. Acad. Sci. **41**:787.)

way to provide the first critical evidence of cell origin in mixed cell aggregates and thus begin the modern analysis of reconstitution. In an extensive series of experiments, in which they mixed dissociated cells of different germ layers of gastrula and neurula stages, they found that cells moved within the aggregate. Ectodermal cells always accumulated at the periphery of the aggregate and endoderm cells in the interior (Fig. 7-4). Mesoderm cells, which were identified by smaller size and absence of pigment, likewise formed homonomic sectors in the interior. Precise cell origin was not established in all instances, since in the case of the mesoderm several tissue types resulted. Nevertheless, these experiments provided conclusive proof that cells in these aggregates tend to segregate according to type and reestablish their former associations. A couple of years later, Moscona found a means of tracing cells within aggregates of chick and mouse cells and showed that here too cells from one organ sort out from those of another. He took advantage of two established facts: cells from mouse and chick embryos are generally distinguishable histologically by their different reactions to hematoxylin stains, and they will differentiate together in culture as if derived from the same species. In an aggregate of dissociated mouse mesonephric cells and chick chondrogenic cells the epithelial tubules are formed by mouse cells and the cartilage by chick cells.

In both of these studies it was utilization of nature's bounties that enabled the investigator to take the first step toward an understanding of reconstitution. All methods have their limitations, however, and it is recognition of their weaknesses as well as their strengths that permits their most effective use. The use of natural markers revealed much about the movements of cells, but they confine one to certain kinds of cells. Moreover, the marker may be lost if the cells undergo transformation. Markers of

EPIDERMIS + MESODERM

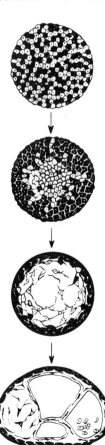

Fig. 7-4 Mixed aggregate of prospective epidermal cells and mesodermal cells from an amphibian neurula, showing sorting out by outward movement of epidermal cells and inward movement of mesodermal cells. The latter eventually forms mesenchyme, coelomic cavities, and blood cells. (After Townes and Holtfreter. 1955. J. Exptl. Zool. **128**:53.)

more general applicability were needed. The best candidates were radioactive isotopes. With isotopes, such as ^{14}C, ^{35}S, and ^{32}P, a means is provided of marking individual cells safely, distinctly, and for sufficient periods of time. For example, by labeling chick mesonephric cells with ^{35}S we were able to show that large numbers of them segregate from chick wing-bud cells in mixed aggregates. However, since ^{35}S was doubtless incorporated into compounds such as proteins or certain mucopolysaccharides, which may turn over or transfer to other cells, we could not be certain that some transfer of label had not occurred. It was at that time that tritiated thymidine became available. I have already discussed, with reference to the neural crest, its sterling qualities as a cell marker (p. 8). When we applied this method to the problem of reconstitution we were able to confirm the previous work and show that chick cells derived from several organs segregate from other kinds of chick cells.

It is clear from these several studies that the affinities demonstrated to be so important at the tissue and organ level are also evident at the cell level. Dissociated cells of germ layers and of organs in relatively advanced stages of differentiation migrate within a mixed aggregate and adhere selectively to cells of like origin. Since studies of cell affinities in advanced stages invariably utilized suspensions of cells derived from whole organs, they of course gave no information on cell specificity at the tissue level. In most organs the individual tissue types are so intermingled that it is virtually impossible to disentangle them and obtain preparations of a single cell type. In the eye, however, one layer, consisting exclusively of cells of one type, can be readily separated from the others. This is the retinal pigment layer or tapetum. Its cells can be easily distinguished in mixed aggregates because they possess black melanin-protein granules. We found that when these cells are dissociated and mixed with cell suspensions from a variety of organs they sort out efficiently. It can therefore be said that for the retinal pigment layer, at least, cell affinity is tissue specific. It would naturally be of great interest to know if tissue affinity operated in the reconstitution of organs—whether, for example, in a differentiated aggregate of dissociated mesonephric cells the epithelial tubules are derived exclusively from former tubule cells, the mesenchyme from former mesenchyme cells, etc. Everyone has assumed that this is what happens. But, in point of fact, evidence for this assumption was largely lacking until recently. Okada has now shown that nephric-tubule cells do indeed sort out from other kinds of cells during the reconstitution of kidney. He obtained his evidence by making fluorescent antibody against nephric-tubule cells. When this antibody was applied to sections of reconstituted kidney, only tubule cells bound the antibody. Other studies like this on other organs are needed.

Even though the picture is incomplete, the phenomenon of sorting out has obvious histogenetic significance. All tissue cells appear to have migratory ability when dissociated and presented with unfamiliar cellular environments or cultured in vitro. They seem to remain immobile only when adherent to their normal cellular associates. The topographic integrity of cellular contacts in tissues and organs, therefore, does not depend on intrinsic inability of the cells to move. The immobility of cells in tissues, and hence the topographic stability of tissues, rests rather on the adhesion of cells to each other (and to the intercellular matrix). The selective nature of this adhesion is obviously a question of first rank developmental and histological importance. Incidentally, the general capacity for cellular mobility revealed by these studies suggests that the transformation of immobile benign tumor cells into mobile malignant cells depends less on the elaboration of a mechanism for movement than on a reduction of cell-to-cell adhesiveness (see pp. 83–84).

Theories of cellular segregation

An adequate theory of sorting out or cellular segregation must explain two aspects of the process: a) the eventual adhesion of cells of the same type to form sectors of like cells and b) the positioning of these sectors within the aggregate in a concentric pattern peculiar to each combination (Figs. 7-3, 7-4). I will only consider proposals that take both of these phenomena into account. It must be emphasized that in all cases two cellular activities are involved—motility and adhesion. Until recently the most popular proposals for sorting out assumed random cell movement followed by qualitatively specific adhesion of cells of like type. This might explain sorting out, but it is difficult to explain positioning within the aggregate on the basis of such a mechanism. On the other hand, any mechanism that accounts for positioning would account equally well for sorting out. Three mechanisms have been proposed to account for positioning. All three assume cells to be indiscriminately adhesive to other cells at first.

Chemotaxis

This hypothesis, promoted mainly by Stefanelli and also at one time by Holtfreter, proposes that dissociated cells respond to differential concentrations of metabolites within the aggregate and migrate along a gradient in concentration either toward or away from the point of highest concentration. The agent (such as CO_2) could be produced by the aggregate as a whole, in which case the highest concentration would be in the center of the aggregate. This would attract the most sensitive cells to the center of the aggregate and cause segregation into an inner and outer component. Certain cell types could also be attracted toward the periphery by a factor in the environment (such as O_2). This would also result in an inner and outer component.

Another form of chemotaxis would involve migration of a particular cell type toward a substance that this type and only this produces. In an entirely dispersed mixture of cells the concentration of this substance would be highest in the middle, and the cells concerned would move toward and accumulate in the middle. Centers of attraction could also form away from the middle if the mixture is not entirely dispersed, i.e., if small clusters of the cell type in question exist within the aggregate at the onset of sorting out. This type of chemotactic action might also occur when a certain number of cells of the same type meet by chance. This might happen at several sites in the aggregate and could lead to the formation of several sectors of cells of one type.

Timing hypothesis

This hypothesis makes the not unreasonable assumption that the cell surface is modified by dissociating agents or by components of the culture medium used for reaggregation. Curtis suggests that cells become migratory as a result of this modification and continue to migrate at random within an aggregate until they recover. As this occurs, their adhesiveness rises. It is assumed that cell movement is promoted by the shearing effects of cells moving over each other. At the surface of the aggregate, cells are in contact over only part of their surfaces, hence there is less shear. Therefore, as cells with increased adhesiveness reach the surface by random movements the combined effect of increased adhesiveness and reduced shear will tend to trap them there. Here they will form an immobile layer, against which other cells of similar type will in turn become trapped. In this manner, all cells of one type will gradually become trapped in the cortical region of the aggregate, and other cells will be herded by them toward the middle of the aggregate. As successive cell types alter their properties, one after another, they will become trapped and build up layers of cells. Thus, both sorting out and positioning of cell types is achieved.

Differential cellular adhesiveness

This hypothesis has been promoted by Steinberg. It assumes that segregation results entirely from random motility and quantitative differences in the general adhesiveness of cells and that an aggregate of cells may be treated as if it were a multiple phase system of immiscible liquids. It further assumes that the thermodynamic theory of liquids would apply, if by analogy the place of the molecules in the physical system is taken by the cells in the biological system. In order to achieve thermodynamic equilibrium there must be maximal adhesion of cell surfaces within an aggregate. If these conditions obtain and the cells are motile, they will tend to exchange weaker for stronger adhesions. The distribution of tissue phases that comes about in a heterogeneous population should then reflect the strengths with which the different kinds of cells adhere to like and unlike cells.

On the basis of such a model, Steinberg has been able to make predictions which could be checked against cell behavior in aggregates. Let us assume that an aggregate is composed of two cell types, A and B, and that A cells cohere more strongly than B cells. If A-B adhesions are intermediate in strength between A-A and B-B adhesions but weaker than the average of the two, a population containing A and B cells will tend to sort out, with a core composed of A cells and the surface occupied exclusively by B cells. Calculations show that these adhesive relationships should obtain if adhesion between two cells depends simply on the frequency of adhesive points on their respective surfaces. Whenever A cells make the smallest contact with other A cells, they will tend to extend the contact. In this way, there will be a continual exchange of heterotypic A-B adhesions for isotypic A-A cohesions, until all, or almost all, A cells cohere. Since cells will tend to cohere maximally over as much of their surfaces as possible, the more adhesive A cells will tend to move internally, away from the surface of the aggregate. The more cohesive A cells will tend also to exclude the less cohesive B cells toward the periphery. By default then, the surface of the aggregate will come to be composed entirely of the more weakly cohering B cells. In short, the more cohesive A cells will form the discontinuous internal component, engulfed by and embedded in a continuous component composed of the less cohesive B cells. Eventually the boundary area of the aggregate will become spherical because the area of contact between the concentric phases will tend to minimize. Since the positioning, according to Steinberg, is due to quantitative differences in adhesiveness, it should vary in different cellular combinations. This was demonstrated to be so by mixing cells from a number of organs, two at a time. Thus, if in a mixture of types A and B, A cells form the cortex and B cells the core; and in a mixture of B and C, B cells form the cortex and C the core; then in a mixture of A and C, A should form the cortex and C the core. This occurs, and in this way one can build up a hierarchy in which those higher on the scale (supposedly less adhesive) always segregate externally to those lower down. Thus, by means of cell motility and quantitative differences in general cellular adhesiveness, both sorting out and positioning may be achieved. Qualitatively specific adhesion is not a necessary postulate.

A critique of the theories of cellular segregation

We are certainly not yet in a position to decide definitively which of these (or other possible hypotheses) provides the proper explanation for positioning. Nevertheless, there are certain lines of indirect evidence that support arguments for and against each.

Under ideal conditions, the first form of chemotaxis would result in the inner component always taking a final position in the center of the aggregate in a single cluster (Fig. 7-5). Actually, the inner component is usually neither central nor single. Nor does the time course of sorting out favor this mechanism. The first small cluster of the internally segregating component are randomly located in aggregates, not located primarily near the center, as they should be if chemotaxis were at work. The facts thus weigh heavily against the primary form of chemotaxis. This is not to say that cells in aggregates may not be attracted along gradients of chemicals produced by cells. The primary form of chemotaxis is simply not a sufficient explanation of sorting out and thus not a useful working hypothesis.

Obtaining evidence that bears on the second form of chemotaxis is more difficult. If the mixture of cell types in the aggregate is evenly mixed at the outset of sorting out, the concentration of hypothetical attractant will be highest in the central region of the aggregate, and the same pattern of sorting out would be expected as for the first type of chemotaxis. However, the original mixture of cells in an aggregate in fact is rarely evenly mixed. Small clusters of two to five cells or so of each cell type may generally be detected here and there within aggregates. Such tiny cell clusters could be foci of attraction and could consequently give rise to some of the results of sorting out that are obtained, i.e., several clusters of the internal segregating component forming at random within the aggregate. In order to obtain critical evidence on the second form of chemotaxis, the movements of individual cells must be observed. If they are chemotactically attracted to cells of the same type, they should move directly toward a cluster of them whenever they are within sensitive range. The only way to obtain information on the direction of movement of individual cells is to observe them directly in living mixed aggregates. Monahan and I did this with time-lapse cinemicrography of individual chick retinal pigment cells sorting out from heart cells. Because of their black pigment granules the pigment cells can be readily followed. The films reveal no consistent pattern of migratory behavior by individual

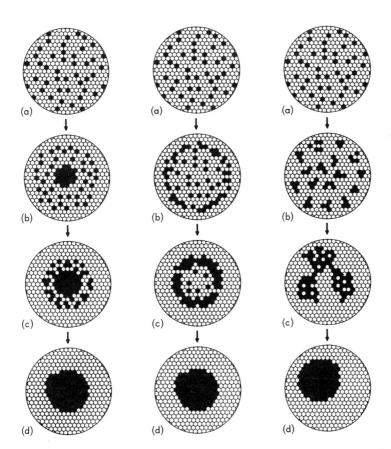

Fig. 7-5 The time course of segregation as it would appear ideally, if brought about through directed migration (chemotaxis), timing, or differential adhesiveness. (Steinberg. 1964. Cell Membranes in Development. Academic Press, N.Y. p. 321.)

pigment cells. They move away from a nearby cluster of pigment cells as frequently as they move toward it. In addition, two pigment cells that have been observed to move toward each other and apparently come into contact may reverse themselves and move apart again. It would seem from these observations that the second form of chemotaxis does not operate in the sorting out of retinal pigment cells from heart cells in the one case in which individual cell behavior has been examined. It is therefore not a good working hypothesis for explaining cell segregation in general.

On the basis of the timing hypothesis Curtis has been able to make certain valid predictions. As we have seen, dissociated cells of an amphibian gastrula sort out with ectoderm at the surface, endoderm in the center, and mesoderm in between. If the surface position of the ectoderm is indeed due to faster recovery from the dissociation treatment, an advantage in time given to the endoderm should reverse the positioning. This is exactly what was found. If endoderm cells are given a head start by dissociating them four hours ahead of the ectomesoderm, the germ layers do not sort out. The cells remain intermingled. If endoderm cells are given a six-hour start, they assume the cortical position, and the ectomesoderm cells move to the interior. In similar experiments with several species of sponges, Curtis found that dissociated cells of different species reaggregate separately at different rates. When cells from two such species are mixed, they sort out. If the slower aggregating cells are dissociated some hours prior to being placed with cells of the other species, however, sorting out does not occur. As expected, when cells from sponges with similar rates of aggregation are mixed, they likewise do not sort out.

All of this is to the good and supports the timing hypothesis. But this is not the whole story. Ideally, if the timing hypothesis is correct, cells of the inner component should be herded passively to the center of the aggregate as the externally segregating cells are trapped at the surface of the aggregate. Thus, at the completion of sorting out the inner component should form a single central cluster of cells. As already pointed out, the inner component is generally neither single nor central. In fact, when the proportion of cells of the inner component is small, they usually sort out very poorly and form clusters located randomly within the reaggregate. Eccentric segregation of the internal component can be explained by the timing hypothesis only if subsidiary assumptions are made, such as variation in the time at which trapping of the external component starts in various parts of the aggregate.

The following evidence is probably even more damaging to the timing hypothesis and at the same time provides support for the differential adhesion hypothesis. It has recently been shown by Steinberg that the same patterns of positioning one obtains in mixed aggregates of two kinds of dissociated cells from chick embryos can be obtained if intact pieces of the same organs or tissues are joined in culture. That which is the external component in a reaggregate spreads over the internal component until a semispherical explant is formed. Holtfreter obtained similar results with intact germ layer explants of amphibian gastrulae and neurulae. Whatever may be the factors that cause one component to move over the other in these experiments, they are not related to recovery from dissociation. It seems most unlikely that peripheral injury from cutting could have differential effects on all the cells.

At this time nothing more can be said. Further studies are now needed to ascertain whether these conflicting interpretations can be resolved. In the meantime, it must be emphasized that the changes in cell adhesion and motility due to shear, postulated by Curtis as part of his hypothesis, still await confirmation. In a word, the timing hypothesis lacks adequate support. It has been useful in calling attention to the largely unanalyzed effects of dissociating agents on the the cell surface, and the effects it postulates may have important modifying effects on cell segregation, but it is inadequate as an explanation of sorting out. Moreover, the timing hypothesis has limited appeal for those concerned with morphogenetic cell movements within organisms, since it requires an artificial effect to explain the tendency of cells to assume normal topographic relationships in mixed aggregates, e.g., mesoderm within ectoderm and cartilage within muscle. (Incidentally, Curtis' experiments with ectoderm and endoderm provide evidence against the operation of a chemotactic mechanism of the first type, since ectoderm can be made to move either inward or outward when mixed with endoderm cells).

The differential adhesiveness hypothesis appears to be the most probable of the hypotheses under consideration. Without need for subsidiary assumptions, it is consistent with and provides an explanation for all features of sorting out that we have described (with the exception of the results of Curtis' experiments). Steinberg has tested the predictions of the differential adhesiveness hypothesis in three ways. First, he showed that when the proportion of the internally segregating cells is small, those at the surface leave the surface as always, but once inside the aggregate they have less chance of encountering other like cells. Hence they tend to remain scattered throughout the interior as individual cells or in small cell clusters. None remains at the surface, but some come to rest just beneath the surface as predicted. Second, he showed that achieving an equilibrium does not depend on cells being previously dissociated. If one takes the intact tissues of a particular combination and joins them in culture, that which is the externally segregating component in the mixed aggregate will spread over the other tissue. The final result will be a spherical explant with the same arrangement of tissues as after sorting out in a mixed aggregate. For example, a chunk of cartilage will be surrounded by a piece of pigmented retina, and heart will be engulfed by embryonic liver. Townes and Holtfreter had previously obtained similar results with intact germ layer explants from amphibian gastrulae and neurulae. Third, since positioning is postulated to be due to quantitative differences in adhesiveness, it should vary in different cellular mixtures. This was shown to occur by mixing cells from several organs in binary combinations. If in separate mixtures A segregates internally to B and B internally to C, one would predict that, in a mixture of A and C, A should form the core and C the cortex. This occurs, and in this way one can construct a hierarchy in which those lower on the scale always segregate externally to those higher up.

Although these observations support the differential adhesiveness hypothesis in a rather impressive way, there are several questions that remain. 1) First and foremost, of course, is whether the postulated quantitative differences in cellular adhesiveness actually exist. Until recently we had no information on this question, nor is it an easy question to answer. As we have seen, the adhesiveness of the cell surface is difficult to measure, and the results seldom lead to undisputed interpretation, especially where small differences in adhesiveness are concerned. Since in the sorting out process we are concerned with the adhesion of cells to other cells, the prefered measurement would of course be of cells to other cells and not to an inanimate substratum. The only measurements germane to the problem therefore are those of Roth and Weston on cells of liver and neural retina (refer to p. 73). It will be recalled that when they counted the number of labeled cells collected by unlabeled aggregates, the isotypic associations were more stable than the heterotypic ones. This demonstrates that differential adhesive stability between these cells exists, and thus in a general way the result is in accord with the differential adhesiveness hypothesis. However, this result is not consistent with one of the important predictions of the Steinberg model. Since liver cells segregate internally to neural retina cells in mixed aggregates, the model predicts that liver-liver associations should be more stable than associations between cells of the neural retina and that heterotypic associations between liver and neural retina cells should be of intermediate stability. The results do not permit a decision as to whether one isotypic association is more stable than the other, but they do show clearly that the heterotypic combinations are always less stable than both the isotypic combinations.

It appears therefore that the differential adhesiveness hypothesis has not withstood its first direct test. This is a severe blow to the hypothesis, but not necessarily decisive. Actually, the results of Roth and Weston give us a choice of two alternative conclusions. Either the differential adhesiveness hypothesis in its present form is inadequate and needs revision, or the probability of forming stable, initial adhesions (as measured by Roth and Weston), giving collision, is not an adequate measure of the adhesive strengths proposed in Steinberg's model. Although we are not in a position to choose between these alternatives at present, I suggest that the hypothesis not be abandoned yet, mainly because it explains so much.

Incidentally, it is not really necessary that adhesion between molecules (in the model liquid system) be represented in mixed aggregates by adhesion between cells. Any cellular behavior that reduces the probability of separation, once cells have collided, will in theory serve instead of division. L. Weiss (see p. 97) has suggested that what keeps cells together may not really be their adhesion, which may be essentially the same and complete in all cases, but the resistance of their membrane to rupture. One could also substitute an inhibition of locomotion on contacting like cells for adhesiveness and get the same results.

Although logic and some experimental evidence favor the thesis that quantitative differences in the surface properties of the cells of different tissues have importance in sorting out and positioning of cells within the aggregate, this does not exclude the possibility of qualitative factors acting as well. Recent evidence from Moscona's laboratory in fact supports this possibility (see p. 95). Such evidence must be taken into account in future evaluation of the various theories of sorting out.

2) Another question concerns the manner of cluster formation by the internally segregating component. It is known from histological analysis of mixed aggregates of chick cells fixed at intervals that the potentially internal cells begin to segregate within a few hours after culturing with the formation of many small homotypic clusters that are randomly distributed. In mixed aggregates of retinal pigment cells and various kinds of other cells this migratory phase is also evidenced by a change in cell form. Individual retinal pigment cells become markedly elongate and dendritic in form (Fig. 7-6). However, as they adhere to other cells of like type they lose the dendritic form and return to a more polyhedral form. Toward the end of the second day in these aggregates, almost all cells become

Fig. 7-6 Chick retinal pigment cell in an aggregate with heart cells. Note two long processes of the pigment cell revealed by the presence of pigment granules. One day in culture. (Trinkaus. Unpublished.)

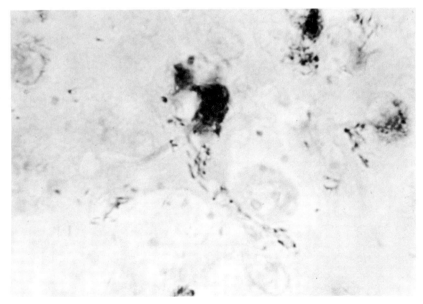

concentrated in a small number of larger clusters but seldom in a single large cluster. It has been pointed out that the frequent segregation of the internal component into two or more randomly located clusters is more readily explained by the differential adhesiveness hypothesis than by the other two hypotheses. But sometimes segregation is very incomplete indeed, ceasing while the internal component is still distributed in several clusters. This raises an important question. If cell clusters as well as individual cells can move and join within the aggregate and if the process of exchange of heterotypic for isotypic adhesions proceeds to completion (as is required by the hypothesis), all cells of the internally segregating component should ultimately come to cohere in a single large internal mass. It is essential therefore to know whether clusters of the internally segregating cells can move.

While on the subject of cluster enlargement, there is still another possible mechanism of cluster enlargement which deserves consideration. Conceivably there could be an exchange of isotypic adhesions for heterotypic ones at one phase of the process, with certain small clusters composed of more weakly adhering isotypic cells tending to disaggregate. The cells thus freed would resume migratory movements within the aggregate and would eventually contact and cohere more firmly with those isotypic clusters that have persisted. By this means the latter would enlarge. If the internally segregating component of a mixed aggregate is indeed the more cohesive one, such a disaggregation of internal clusters should not occur.

3) Both the first form of chemotaxis and the timing hypotheses require considerable migratory ability on the part of some of the cells. (The second form of chemotaxis does not necessarily require extensive migration.) The differential adhesiveness hypothesis, on the contrary, does not require cell migrations over long distances. There are only two reasons for movement to occur. a) Cells that are more adhesive will tend to leave the surface of the aggregate. b) There will also be an exchange of weaker heterotypic adhesions for stronger isotypic ones. Cells will tend to spread maximally over adjacent cells in order to have maximum adhesive contact. This would increase the probability of their contacting isotypic cells. By spreading over heterotypic cells they contact in order to cohere with them maximally, they may take on dendritic shapes (as indeed is the case for retinal pigment cells during the migratory phase). As isotypic cells cohere and pack together, on the contrary, they will tend to lose their extended condition and take on a more polyhedral form. Retinal pigment cells do this when they contact each other. If internally segregating cells do not have isotypic cells in the vicinity and thus fail to contact them, they may exhibit no translocation and sorting out. Under such conditions they will simply exercise in place. Since chemotaxis is apparently not at work, they have no way of knowing if other isotypic cells are in the vicinity, unless they

contact them. In such a situation, they might retain the spindle or dendritic shape. In this connection, it is interesting to note that individual retinal pigment cells that have failed to find a mate tend to remain dendritic. If the movements of the internally segregating cells are truly limited, it may be predicted that the lower the proportion of internally segregating to externally segregating cells, the poorer will be the sorting out. Indeed, at very low proportions there should be essentially no segregation but simply a movement of the internally segregating cells away from the surface of the aggregate. If either the first form of chemotaxis or the timing hypothesis holds, the internally segregating cells should always move (or be moved) toward the center of the aggregate, no matter what their proportion in the aggregate.

It is evident that detailed information on the movements of cells and cell clusters within the aggregate would answer some critical questions and provide a further test of the hypotheses. To gain this detailed information, it was necessary to observe segregation directly. All conclusions reached up to this point have a strong element of extrapolation in them, because they are based on histological observations of different aggregates fixed at graded intervals. The only way to determine what is in fact the behavior of cells and cell clusters during the process of sorting out is to study the entire process at closely timed intervals in individual living cell aggregates, in which the internally segregating cells are visually distinguishable. Moreover, the aggregates must be three-dimensional systems in which sorting out occurs in an exclusively cellular environment, rather than on an inert substratum such as glass. For these various reasons, Judith Lentz and I decided to make such a study, choosing as our system mixed aggregates of retinal pigment cells and heart cells of the chick embryo. In such a combination, the pigment cells constitute the internally segregating component and are readily visible because of their melanin-protein granules. We varied the proportions of the two cell types and followed photographically, at closely timed intervals, the entire process of sorting out (Fig. 7-7).

The onset of sorting out of retinal pigment cells is already evident within a few hours after aggregation. It continues steadily and is largely complete in 48 hours. During this phase, movements of individual cells appear to take place only over very short distances (10–30 μ). The movements also appear to be random, except of course for those cells which were initially at the surface of the aggregate. These move internally. Sorting out is less complete when the proportion of pigment cells to heart cells is less than 1 : 4 and hardly occurs at all when it is reduced to 1 : 16.

With a sufficient proportion of pigment cells, small clusters form at random internally within a few hours. No clusters have been observed to move, except those so tiny as to be just distinguishable from individual

120 *Cellular Segregation and Selective Adhesion*

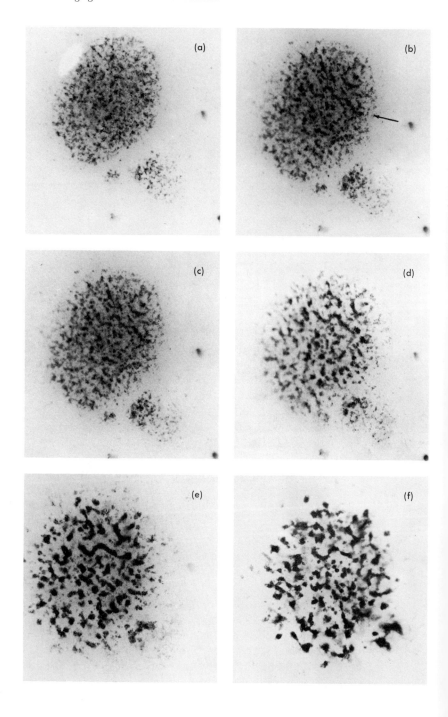

cells. Clusters enlarge by accretion, the adhesive addition of individual cells and tiny clusters. As clusters enlarge, they often contact adjacent clusters and fuse. Clusters are constantly changing shape, presumably because of cell movements within each cluster, as isotypic cells spread over each other. These form changes often cause adjacent clusters to contact each other and fuse. Once small clusters of pigment cells have formed, they persist, and it is possible to trace them from hour to hour and day to day, unless they lose their identity by fusing with adjacent clusters. There is no sign of disaggregation into still smaller clusters or individual cells. One would predict from these observations that the higher the proportion of pigment cells, the smaller the number, and the larger the size of the clusters they form. This is in fact the case. During the final phases of sorting out (after approximately 48 hours) the clusters which have increased in size due to accretion and fusion with other clusters now round up and become more compact, as expected of cells cohering maximally. There is also a contraction of networks of interconnected clusters of pigment cells to form more compact cell masses.

These results clearly favor the selective adhesiveness hypothesis, at least in so far as the sorting out of retinal pigment cells from heart cells is concerned. Pigment cells move internally over very short distances and, as predicted, sort out poorly when the proportion of them is low. Clusters of pigment cells do not disaggregate, do not move, and fuse only when brought into contact by accretion or form changes. This explains why the smaller the proportion of pigment cells (in the range where sorting out occurs), the smaller the size, and the larger the number of the clusters they form. In view of all this, and even taking into account the contradictory result of Roth and Weston, it seems reasonable to accept the selective adhesiveness hypothesis as the most likely mechanism of sorting out in mixed cell aggregates.

There are, however, still other questions. This is by no means a discouraging situation, for as will become apparent, the questions that we may now begin to ask are largely direct outgrowths of research of the last few years. They therefore represent progress.

The most immediate question is why clusters of retinal pigment cells do not move within the aggregate. This question is particularly pertinent

---

Fig. 7-7 Mixed aggregate of $5\frac{1}{2}$-day chick retinal pigment cells and 4-day heart cells in a ratio of 1:4, showing the formation of small clusters of pigment cells and their gradual fusion to form larger clusters. Note in particular the fusion of several randomly arranged clusters to form a single, large, elongate cluster [arrow in (b)]. Clusters do not move. In (e) and (f), clusters are condensing. This causes some clusters to pull apart (see text). (a) 12 hours in culture. (b) 20 hours in culture. (c) 23 hours in culture. (d) 34 hours in culture. (e) 68 hours in culture. (f) 5 days in culture. (Trinkaus and Lentz. 1964. Develop. Biol. 9:115.)

in view of the fact that the sponge workers have observed that small clusters of sponge cells may move. It may be objected that this movement of sponge cell clusters occurs on a glass substratum and therefore does not apply because of the different conditions. This is a valid objection. It is therefore of particular interest that it has been shown recently that clusters of endocardial cells within the early chick blastoderm translocate within an exclusively cellular environment. It is not known at present why clusters of pigment cells do not move in our preparations, but it seems likely that this lack of movement is related to the small amount of translocation exhibited by individual pigment cells. Where cluster movement occurs, it is probably due to the collective movements of the constituent cells. We have directly observed this to occur in dissociated cells of a teleost blastula. The movements of these cells are so active that when several are clustered together one can readily observe the translation of the sum of individual cell movements into movements of the group. It would be of interest to know if the heart cell clusters in locomotion within the chick blastoderm are composed of cells that individually have high locomotory capacity (see Ch. 13).

Chick retinal pigment cells appear to have such low mobility that it seems possible that some of them do not actually translocate at all but merely exercise in place, sending out and withdrawing long filopodia. But if these filopodia contact a more adhesive surface, such as that of another pigment cell, they may stick, and then, by contraction of the filopodium, the two cells may be drawn together. If this extremely low level of motility applies generally, it may be necessary to substitute changes in cell shape, such as assumption of dendritic form and contraction of filopodia, for the motility component in Steinberg's scheme.

Although chick retinal cells have low migratory powers when mixed with heart cells, it would be premature to generalize. They might migrate more actively if mixed with other cells. Also, other kinds of cells may have greater migratory powers than they. For example, the heart cells in our preparations could well have extensive migratory activity; but since they were not marked their movements could not be followed. Similar studies are obviously needed on other kinds of cells. The problem will be how to follow them in the living state. A marker that can be readily detected, like melanin pigment, will be needed.

There are certain similarities between the phenomena of contact inhibition and sorting out. In contact inhibition, when a fibroblast contacts another fibroblast moving on glass its ruffled membrane ceases its activity at the point of contact, and the cell stops moving. Since it has been thought that cells cease moving during sorting out when they contact isotypic cells, it has been suggested that contact inhibition is also at work here. Curtis used this as a partial basis for his hypothesis of segregation. Because of the ease of studying cell behavior at the periphery of a monolayer on a transparent inert substratum the possible similarity of contact inhibition and sorting out has much experimental interest. If contact inhibition operates during sorting out, cells should cease moving or almost so when they contact cells of like type. We have already seen that once retinal pigment cells contact isotypic cells they essentially cease locomotion. It may be objected that the clusters of pigment cells change shape somewhat and that this is evidence of a certain amount of cell surface activity or movement within the cluster. This is indeed probable. However, it does not vitiate the suggestion that contact inhibition may be at work. Even under the classical conditions for studying contact inhibition in tissue culture the inhibition of movement is never complete. The same may be said for the time-lapse observation of P. Weiss and Taylor that some cell movement continues in isotypic aggregates of liver, lung, and kidney cells in culture. A more direct approach has been to label intact pieces of embryonic tissue with tritiated thymidine and fuse them to unlabeled pieces of the same or different tissues in culture. Under these conditions, the line separating the labeled from the unlabeled cells remains distinct. There is apparently no cell movement. These observations are all consistent with the proposal that the formation of isotypic cell clusters during sorting out in mixed aggregates is due in part to the operation of contact inhibition.

But this is not the whole story. We have not taken into account another characteristic of contact-inhibiting cells in tissue culture. There is a distinct tendency for cells at the periphery of the sheet to break away from the other cells and move out on the inanimate substratum. Cells which have sorted out into isotypic clusters do not behave this way. Individual retinal pigment cells have not been observed to leave pigment cell clusters; cells do not move away from isotypic aggregates of liver, lung, and kidney epithelial cells in culture; and there is no mutual invasion by labeled and unlabeled cells in fused explants. The reason for this difference could be that the choice before the cells in the different systems is not the same: cells versus inanimate substratum as opposed to cells of one kind versus cells of another kind. If Carter is correct in his proposal that contact inhibition of cells in sheets is due to their greater adhesiveness for the inanimate substratum than for each other (see above, p. 25), then perhaps cells do not leave isotypic clusters within aggregates because they are more adhesive to cells of the same type than to the other kinds of cells which are present. One way of getting further information on this interesting question would be to observe particular combinations of contact-inhibiting and noncontact-inhibiting cells in both experimental situations.

While on the subject of contact inhibition, it should be noted that if certain kinds of dissociated cells are placed on a glass substratum, they will often move around until they make contact with other cells. Then, instead of being inhibited in their movements, they move preferentially over each other and reunite into adherent groups or aggregations in which individual cells continue to move about. This suggests that certain cells are not contact inhibiting, at least under certain conditions. The phenomenon may be due to greater adhesion to other cells than to glass, but in reality it is not all understood. For convenience Curtis called it "contact promotion."

A question of enduring histogenetic interest is whether cells derived from tissues of the same organ sort out from each other. According to the differential adhesiveness hypothesis, there must be differences in adhesiveness between the different tissues cells of an organ during organogenesis. How else would they have become arranged that way within the organ? As already pointed out, experiments on this problem will not be easy to perform, but we already have a model experiment in the fluorescence-antibody labeling of cells by Okada (see p. 108).

It is perfectly clear that, regardless of how they operate to control cell positioning, both cell motility and differential adhesiveness are of critical importance not only within artificial aggregates of cells in culture but during normal development as well. It is therefore of interest to note that apparently neither of these properties is fixed in embryonic cells. In amphibian embryos, Jones and Elsdale have shown that the adhesive and aggregative properties change markedly in cells from blastula, gastrula, and neurula stages. We have also observed these properties to change from the blastula to the gastrula stage in fish embryos. And Kuroda and Moscona have found that disaggregated cells from chick embryos of different ages show changing patterns of aggregation in the rotary shaker. The detailed working out of these changes in cell behavior as they relate to discrete developmental events has only just begun. It has not yet been determined, for example, whether these changes affect sorting out and if so how. The possible developmental consequences of such changes will be discussed when we consider mechanisms of gastrulation.

SELECTED REFERENCES

Chiakulas, J. J. 1952. The role of tissue specificity in the healing of epithelial wounds. J. Exptl. Zool. 121:383–418.

Curtis, A. S. G. 1960. Cell contacts: some physical considerations. Am. Naturalist. 94:37–56.

Curtis, A. S. G. 1961. Timing mechanisms in the specific adhesions of cells. Exptl. Cell Res. Suppl. 8:107–122.

Galtsoff, P. S. 1925. Regeneration after dissociation. (An experimental study on sponges.) I. Behavior of dissociated cells of *Microciona prolifera* under normal and altered conditions. J. Exptl. Zool. **42**:183-222.

Hilfer, S. R. and E. K. Hilfer. 1966. Effects of dissociating agents on the fine structure of embryonic chick thyroid cells. J. Morphol. **119**:217-231.

Holtfreter, J. 1939. Gewäbeaffinität, ein Mittel der embryonalen Formbildung. Arch. Exptl. Zellforsch. **23**:169-209.

Moscona, A. A. 1952. Cell suspensions from organ rudiments of chick embryos. Exptl. Cell Res. **3**:536-539.

Moscona, A. A. 1957. Development in vitro of chimaeric aggregates of dissociated embryonic chick and mouse cells. Proc. Natl. Acad. Sci. U.S. **43**:184-194.

Okada, T. S. 1965. Immunohistological studies on the reconstitution of nephric tubules from dissociated cells. J. Embryol. Exptl. Morphol. **13**:299-307.

Sperry, R. W. 1965. Embryogenesis of behavioral nerve nets, p. 161-186. *In* R. L. DeHaan and H. Ursprung [ed.] Organogenesis. Holt, Rinehart & Winston, New York.

Steinberg, M. S. 1963. Reconstruction of tissues by dissociated cells. Science. **141**:401-408.

Townes, P. L. and J. Holtfreter. 1955. Directed movements and selective adhesion of embryonic amphibian cells. J. Exptl. Zool. **128**:53-118.

Trinkaus, J. P. 1965. Mechanisms of morphogenetic movements, p. 55-104. *In* R. L. DeHaan and H. Ursprung [ed.] Organogenesis. Holt, Rinehart & Winston, New York.

Trinkaus, J. P. and J. P. Lentz. 1964. Direct observation of type-specific segregation in mixed cell aggregates. Develop. Biol. **9**:115-136.

EIGHT

Movements of Cell Sheets

Until now we have concerned ourselves almost exclusively with the activities of individual cells, moving as separate individuals in the organism, in cell aggregates, and in culture. But in fact the bulk of the movements that mold the embryo during morphogenesis are not movements of individual cells. The principal movements in gastrulation, neurulation, and in the formation of many organs involve cell sheets. Whether it be cells flowing over the lips of the blastopore, or spreading down from the animal hemisphere, or invaginating to form an archenteron or a lens vesicle, or rolling up to form a neural tube, it is cell sheets that do the work and form the structures. Although our digressions are often long, the effort of this book, and indeed of developmental biology, is to understand development the way the organism does it. Eventually, then, we must return to the organism and try to fathom how it moves sheets of cells around and constructs organs out of them. Since our essential concern is with cohesive sheets of cells and since these have been studied in tissue culture and wound healing, we will turn first to these to see what ideas can be extracted.

The spreading of epithelial sheets

Epithelial cells characteristically spread as a flat cohesive sheet in culture. They will also move through a three-dimensional gel in very close association, in the form of cords and bands. It is possible that mutual cellular adhesion is reduced during spreading, as compared to the stationary state, but it is not evident. Cells do not need to break away from their neighbors in order to move, and cells within the sheet neither climb over each other nor lag behind as the sheet spreads. An epithelial sheet will invariably spread, whether in vitro or in vivo, provided it adheres to a substratum and possesses a free edge. Thus, in normal wound healing the first detected cellular activity is the spreading of the epithelium to close the wound. This phenomenon has been observed so many times under so many conditions that one says "an epithelium will not tolerate a free edge." Accompanying the advance of the free edge and thoroughly integrated with it is a similar advance of the sheet behind it. (Of course, many epithelial sheets which spread during embryogenesis do so in the absence of a free edge. This kind of spreading is poorly understood and will be considered in connection with the mechanisms of morphogenetic movements.)

The prime movers in the spreading of an epithelial sheet appear to be the marginal cells. Three lines of evidence support this conclusion. 1) A spreading epithelial sheet is attached to the substratum most firmly, or even only, at its margin; when the margin of a sheet is detached from the glass the whole sheet usually retracts. Since adhesion to the substratum is necessary for the spreading of cells, it would appear that the most active spreading region is at the margin. 2) At the onset of spreading of a wound epithelium only the marginal cells show movement. Later, nonmarginal cells progressively join the advance. 3) The free edges of the marginal cells show marked ruffled membrane activity.

The spreading activity of the marginal cells raises an important question. Are the submarginal cells only passively pulled, or are they active participants? We know that all cells of a sheet are capable of forming ruffled membranes, because when a sheet is cut it will readhere at its new margin and the new marginal cells will form ruffled membranes and begin spreading. Vaughan and I investigated this problem further with time-lapse cinemicrography of spreading epithelial sheets in culture and found that when local portions of the margin spread more than the rest, the cells between such outgrowths become markedly stretched and tangentially oriented. In addition, the submarginal cells of the outgrowth region become radially oriented with respect to the main cell mass (Fig. 8-1). This is clear evidence that cells lacking ruffled membranes, whether marginal or submarginal, respond passively to the pull of the outward spreading membrane-active parts of the margin.

Yet even though cells of the sheets are usually closely adherent to each other, they frequently appear to be pulled apart, as if tension from the spreading margin is too great. When this happens, there is a local retraction

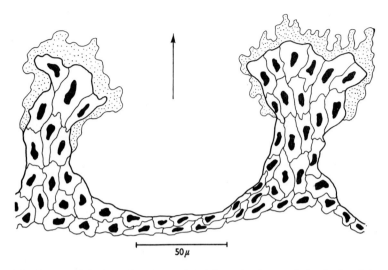

Fig. 8-1 Drawing of the free edge of an epithelial sheet of embryonic chick cells. Stippled areas indicate where ruffled membrane activity was seen. The arrow shows direction of movement of the whole sheet. Two large outgrowths are seen which show membrane activity at their tips and radial orientation of their constituent cells. Tangentially arranged cells are seen between them. No membrane activity is seen in these cells. (Vaughan and Trinkaus. 1966. J. Cell Sci. 1:407.)

and a large gap opens up, often as large as one or two cells. After a brief pause, ruffling begins at the free edges of the gap; and the cells move together, closing the gap. From these observations, it appears that the submarginal cells of a sheet both stretch passively and spread actively. They are stretched passively by the actively spreading margin, and they spread actively intermittently to close gaps that occasionally open up between them due to tension in the sheet. They appear to spread actively only when confronted with an adjacent space.

It has been known for a long time that the spreading movement of such a sheet is stopped by contact with another epithelial edge. This cessation of movement on contacting another epithelium must certainly be a form of contact inhibition. In fact, it has been shown that the edge of an epithelial outgrowth in culture displays much contact inhibition to other epithelia and even to fibroblasts. If spreading of the sheet is due to the collective activity of its constituent cells, perhaps contact inhibition also acts to prevent overlap of these cells and thus coordinates their spreading movements. This was confirmed in the time-lapse films. When the contacts of cells within an epithelial sheet are broken, ruffled membranes form at the free cell borders, and the cells move to close the gap. However, when the cells contact each other again, the ruffled membrane activity stops and with it the spreading of the cells. In the case of fibroblasts, which have been studied in more detail and which show classical contact inhibition of cell movement, spreading of a cell sheet in culture may be completely accounted for by contact inhibition. The more neighboring cells a fibroblast adheres to, the slower its speed of locomotion. But along with this reduced speed there is an improved consistency of movement so that a given cell tends to move in sequence with the rest of the sheet. The net displacement during a given period of time will be much greater for a coherent fibroblast sheet than for isolated cells. Contact inhibition accomplishes this in four ways. 1) Contact inhibition with the cell behind prevents a given cell from reversing its direction of movement and continually stimulates it to move. If it moves too fast, the loss of contact causes it to stop until the ones behind have caught up. 2) Contact inhibition of movement prevents lateral overlap, thus orienting cells in the sheet. 3) Contact inhibition with cells in front prevents a cell from moving forward too far and too fast. 4) Contact inhibition will reduce greatly the mobility of a cell, but if the cell to which it adheres is moving it will tend to move with it. As Abercrombie has emphasized, "For a cell to become anchored it must adhere to something which does not itself move." This could explain the abrupt cessation of movement of cells back of the edge when the marginal cells stop, for now the first line of nonmarginal cells would be anchored to something which does not itself move. The cessation of movement in a contact-inhibiting situation would then be propagated serially, away from the margin of the sheet.

Although epithelial cells appear to be contact inhibiting, they are different from fibroblasts in that they show less tendency to break away from the sheet and move about as individuals. Presumably their adhesive contacts are firmer. It is also possible that they have less efficient ruffled membranes. Direct measurement of cell-to-cell adhesiveness and comparison of fibroblast-fibroblast and epithelial cell–epithelial cell contacts at the level of fine structure should help explain this difference.

In the meantime, there are a few observations of cell activity within small cell clusters that give support to the view that even though a sheet spreads as a unit it is the activities of its constituent cells that provide the momentum. Galtsoff and Holtfreter have seen a coordination of the movements of individual cells within small groups of sponge cells and amphibian embryonic cells that causes the group to move as a unit, and I have observed something similar in small aggregates of dissociated blastomeres from *Fundulus* blastulae. When the cohering cells are strung out in a chain the lobopodial activities of the individual cells can be easily observed. In such a situation the collective movements of the pseudopods are translated into group activity, and the entire chain writhes and twists as a unit. In each of these examples there is no net directionality to the movement. The entire edge of each aggregate is free, and movement is apparently random, whether it be translocation over the substratum or exercising in place. It would be instructive to have some information on cell activity in groups whose movement has a directional quality. Perhaps clusters of presumptive heart cells in the chick blastoderm would provide good material for such a study. DeHaan has shown that they move directionally from their point of origin in the extraembryonic mesoderm to their proper place within the embryo. Some additional support for the dependence of directional group activity on the activity of the constituent cells comes from an unusual source, cells which ordinarily operate as completely separate individuals. Spermatozoa and spirochetes sometimes line up in groups with their axes parallel. When this happens they beat in synchrony and move directionally. G.I. Taylor has shown that the energy dissipation is lowest when they beat together, an observation of undoubted significance; it may give synchrony a selective advantage over asynchrony.

With our emphasis on the adhesion of cells to each other and to the substratum as their collective motility provides a possible mechanism of sheet movement, we have ignored an important variable. To be sure, a cell must have a grip on its substratum in order to move. But its ability to move and its rate of movement rest on balance between its adhesion and its impulse to move. One kind of cell will move more or less than another kind under similar conditions. Hence, when cellular sheets spread at different rates or cease to spread, we must always ask which is primarily involved—a change in cell-cell and cell-substratum adhesion or a change in cell motility.

SELECTED REFERENCES

Abercrombie, M. 1961. The bases of the locomotory behavior of fibroblasts. Exptl. Cell Res. Suppl. **8**:188–198.

Abercrombie, M. and C. A. Middleton. 1968. Epithelial-mesenchymal interactions affecting locomotion of cells in culture, p. 56–63. In R. Fleischmajor and R. E. Billingham [ed.] Epithelial-mesenchymal interactions. Williams and Wilkins, Baltimore.

Holmes, S. J. 1914. The behavior of the epidermis of amphibians when cultured outside the body. J. Exptl. Zool. **17**:281–295.

Lash, J. W. 1955. Studies on wound closure in urodeles. J. Exptl. Zool. **128**:13–28.

Lewis, W. H. 1922. The adhesive quality of cells. Anat. Record. **23**:387–392.

Vaughan, R. B. and J. P. Trinkaus. 1966. Movements of epithelial cell sheets in vitro. J. Cell Sci. 1:407–413.

Wilbanks, G. D. and R. M. Richart. 1966. The in vitro-interaction of intra-epithelial neoplasia, normal epithelium, and fibroblasts from the adult human uterine cervix. Cancer Res. 26:1641–1647.

NINE

Some Guiding Principles in the Study of Morphogenetic Movements

Gastrulation is a process of early development during which the regions of the many-celled blastula are drastically rearranged, so that at its completion the cells that will eventually form each organ system are all in their proper places. In short, it is the means utilized by the embryo to lay down its basic body plan. Since the first glimmer of its significance was perceived, it has entranced all who concern themselves with morphogenesis. For, along with its consequence for differentiation, gastrulation combines cell movements which in their aggregate are the most sweeping to occur during all of embryogenesis.

Although gastrulation may be conveniently divided into a number of particular movements for convenience of analysis, it is essentially a phenomenon of the whole in which the entire embryo is ultimately involved. Each of the movements depends directly or indirectly on every other. Its cardinal feature is integration. For this reason it is the process par excellence in which it will ultimately be necessary to understand each movement in relation to the others in order to have a really meaningful comprehension of each one separately. I will therefore not attempt a horizontal approach in which each constituent movement, such as involution or epiboly, is considered across the board in the various eggs in which it occurs. In my opinion, such a comparative approach is not at present the most profitable. The temporal sequence and geographical extent of the different movements vary greatly from one phylum or class to another. At this point, it is best to consider each organism as a special case whose particular characteristics must be appreciated. Accordingly, I propose to consider, each in its turn, the several organisms whose morphogenetic movements have been subjected to analysis. Hopefully, certain common features that may apply generally will emerge from the presentation.

Before we move further, it will be useful to compile a list of the kinds of questions to be asked. These arise mainly from our knowledge of cell sheets in general and are to be kept in mind as we seek mechanisms.

a) Is the spreading due to intrinsic factors or to an external pulling force? Is the spreading of the cell sheet due to forces within the sheet, or is it stretched by a contractile structure attached to its periphery? In the whole egg we must take account of the fact that each cell sheet is attached to other membranes. If one of these contracts or invaginates it could well exert a powerful pulling force on the layers with which it is confluent. A priori, this is a possibility in epiboly of the ectoderm in an amphibian egg (attached to involuting mesoderm), epiboly of the blastoderm and periblast in a teleost egg (attached to the contractile yolk cytoplasmic layer), and invagination of the archenteron of an echinoderm egg (attached to the contractile filopodia of the secondary mesenchyme).

b) Is there active spreading of the whole or passive stretching by an actively moving leading edge? This possibility has already been considered in the discussion of the spreading of epithelial sheets in culture.

c) What is the unit of activity: the sheet or the individual cells of which it is composed? In a sense, this is a philosophical problem involving the significance of levels of organization. Do the parts lose themselves in the whole and acquire new characteristics that are totally dependent on their associations within the whole? To put it more concretely, does the spreading activity of the sheet depend primarily on the independent spreading activities of the cells, or is the spreading of the cells dependent on their associations within the sheet?

An indication that this is a scientifically valid question comes from a well-known experiment performed at Woods Hole at the turn of the century. F. R. Lillie found that if eggs of the marine annelid *Chaetopterus* are exposed to seawater containing an augmented concentration of KCl before or at the time of fertilization, the eggs are parthenogenetically activated, but cleavage is suppressed. What is remarkable is that some of the events of early development take place anyway. There is a separation of various types of cytoplasm into different regions, a localization of vacuoles in the midprototrochal region, and a differentiation of cilia. As Lillie said, "This structure possesses certain undeniable resemblances to a trochophore." What is of especial interest in the present context is that the outermost clear cytoplasm moves down over the vegetal material as in normal development and recalls the epiboly of cells of the animal hemisphere that occurs during gastrulation. Inasmuch as this occurs in the absence of cellularization, it is clear evidence that the movement of epiboly in these eggs may be at least in part independent of cells. A change occurs in the animal cytoplasm at the appropriate time which causes it to spread as a unit, regardless of whether it is cellular or not. This does not necessarily imply that this change might not occur in each animal ectodermal cell independently of the others during normal development. What it does indicate is that cellularization is not completely necessary. The cytoplasmic sheet may act as the unit.

Holtfreter has observed that a somewhat parallel phenomenon results in unfertilized amphibian eggs if they are left in Ringer solution for two to three days. Uncleaved pigmented cytoplasm of the egg cortex moves down from the animal hemisphere and invaginates like the gray crescent in the upper part of the vegetal hemisphere. Again, movements of gastrulation appear to be partly simulated by uncleaved cortical cytoplasm. It should be emphasized that, extraordinary though these movements may be, they are only partial imitations of the normal. It would appear that cells are necessary for really proper execution.

An obvious example of the movement of a sheet of cytoplasm in the absence of cellularization is the flow of the cortical cytoplasm of an activated teleost egg to the animal pole, where it accumulates to form the blastodisc. This is of course precisely the opposite of spreading and is mentioned here only to illustrate the capacity of a layer of uncleaved cytoplasm to move great distances in a highly controlled manner. In the same teleost eggs there is another perfectly normal movement that is very much to the point. The syncytial periblast, which originates beneath the blastoderm prior to gastrulation, spreads over the yolk to encompass it entirely during gastrulation, without apparent increase in mass. No better example is available to illustrate the capacity of a noncellular protoplasmic layer to spread extensively. Sometimes the uncleaved teleost blastodisc can also be induced to do this. Trifonowa found that some parthenogenetically stimulated eggs do not cleave but nevertheless undergo epiboly. The whole uncleaved blastodisc spreads over the yolk to enclose it completely.

d) When spreading is due primarily to activities of the constituent cells, is there a change in surface properties of the cells, such as adhesiveness or motility? In addition, does the change in cellular activity occur as a result of a change in the cells or in the character of their substratum?

e) Are the changes in cell form that occur during the spreading and folding of cell sheets accompanied by changes in the number and orientation of cytoplasmic organelles, such as microtubules and microfilaments?

f) How are the changes in the behavior of particular cell sheets (from whatever cause) integrated with the events occurring elsewhere in the egg, so that gastrulation and other morphogenetic movements proceed normally?

TEN

Echinoderm Invagination

Many echinoderm eggs are superb material for studying cell movements because of their striking transparency, small number of cells (1000–2000), rapid development, hardiness, and availability for investigation in many parts of the world during long spawning seasons. For the study of gastrulation these eggs have the additional feature of forming the archenteron by a classical, textbook invagination of a cell sheet one cell layer thick. The vegetal plate indents directly into the blastocoel, accompanied by little or no involution (turning over the margin of the blastopore) and little growth or cell division (Figs. 10–1, 10–2). It is an excellent example of both a change in curvature and an extensive increase in the area of a cell sheet. The process continues until the tip of the archenteron reaches the animal pole and is complete when the tip turns ventrally to come to rest opposite the future mouth region, with which it eventually fuses. Laboratories of three continents have taken advantage of the favorable features of these eggs to study several aspects of early development, including the mechanism of invagination.

The modern attack on the problem began on the Pacific coast of the United States in the thirties with the observation of Moore and Burt that the first phase of invagination by the vegetal plate is not dependent on the rest of the egg for its execution. They cut eggs of the sand dollar *Dendraster* into vegetal and animal halves and observed that the vegetal halves began invagination regardless. The indentation was small, about $\frac{1}{4}$ to $\frac{1}{3}$ the distance toward the animal pole. It appeared from this that

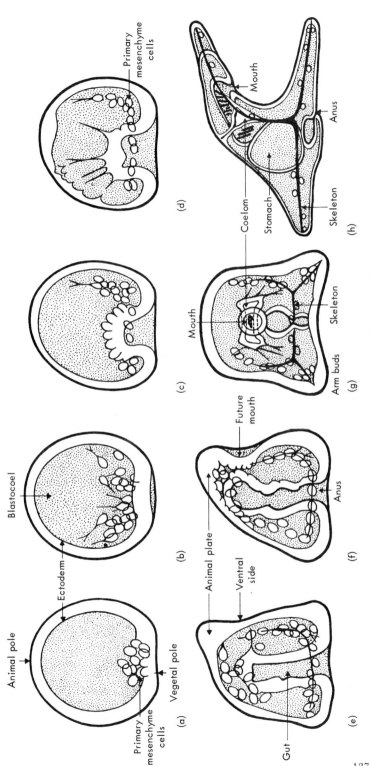

Fig. 10-1 Some main stages in the early development of the sea urchin embryo. The primary mesenchyme leaves the vegetal plate to enter the blastocoel in (a) and (b). The first phase of invagination is shown in (c) and the second phase starts in (d). Note how the primary mesenchyme cells have become arranged in a necklace encircling the archenteron. Filopodia of the secondary mesenchyme project from the tip of the archenteron. In (e) and (f) the flattening of the wall on the ventral side is evident, as is the sharp curvature and thickening at the animal plate; (g) is viewed from the ventral side; (h) is the pluteus larva. The change from (f) to (h) is associated with growth of the skeleton. (Wolpert and Gustafson. 1967. Endeavour. **26**:85.)

138 *Echinoderm Invagination*

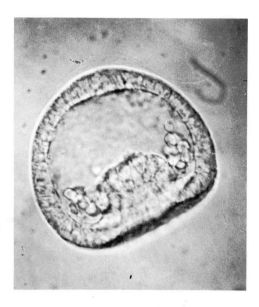

Fig. 10-2 Early gastrula of the sea urchin, *Paracentrotus lividus*. The archenteron has invaginated $\frac{1}{3}$ of the way toward the animal pole. The loose cells on either side of it are primary mesenchyme. ×60. (Trinkaus. Unpublished.)

forces intrinsic to the vegetal plate (which forms the archenteron) are responsible for beginning invagination in these eggs. This experiment also rules out the possibility that differences in hydrostatic pressure between the blastocoel and the outside play a role in invagination (unless, of course, the effect has taken place prior to cutting away the animal half of the egg).

With the interruption of the war, the subject rested there without further attention until the 1950s. In 1956 two rather remarkable papers appeared, reporting work on invagination that had gone on simultaneously and independently in widely separated laboratories. Dan and Okazaki worked on *Clypeaster, Mespilia,* and *Pseudocentrotus,* which are found in the Sea of Japan, and Gustafson, at first with Kinnander and later with Wolpert, used the eggs of *Psammechinus miliaris* from the Kattegat. Both groups concerned themselves initially with the second phase of gastrulation, after the secondary mesenchyme cells had appeared, but Gustafson and Wolpert have also illuminated the primary phase. They followed gastrulation with the aid of time-lapse cinemicrography.

The films show that during the first phase of invagination the columnar cells of the vegetal plate have reduced contact with one another and a rounding up of their inner borders but they retain full contact with the hyaline plasma layer at the egg surface. Their rounded inner borders show pulsating activity. Gustafson and Wolpert point out that if contact between columnar cells is reduced, the change in the packing of the cells would require that the sheet increase in area. They postulate that the vegetal plate is confined by a ring of increased cellular contact of the lateral blastocoel wall and therefore cannot spread as a flat sheet. The alternative is to curve. The direction of invagination will be toward the side where the cells have reduced contact with each other. In the sea urchin vegetal plate this is the inner surface, so the curvature is inward and represents the beginning of invagination.

This concept may well be correct, but it needs scrutiny. The Moore-Burt experiment demonstrates that the blastocoel wall need not be intact for invagination to begin and raises questions about the postulated confining role of the wall of the blastocoel. Because of this it would be of interest to see if the activities of the vegetal plate cells are similar in isolated vegetal halves and intact eggs. The lack of pulsatory activity on the outer surface of the vegetal plate appears to be due to adhesion of cells to each other by means of septate desmosomes and to the hyaline plasma layer. No pulsations are seen where the surface of the cells is in contact with the hyaline plasma layer, but such activity appears at once when the hyaline layer is weakened by treatment with seawater lacking calcium and magnesium. The reduced contact between cells at their inner surfaces and their pulsatory activity are both thought to be due to decreased intercellular adhesion. But there is no direct evidence on this. Moreover, a similar result may follow from other causes, such as an active rounding up of the cells due to an increased tension in their membranes, so that they conform less readily to the contours of adjacent cells. For example, the contact between cells in the sea urchin blastula is reduced when there is an increase in tension at the surface during cell cleavage. It also may be asked to what extent the beginning of active locomotory activity of the cells contributes to their change in form. It should be pointed out that just before and during this process some cells leave the vegetal plate to become the primary mesenchyme (p. 32–33). This is undoubtedly an extreme aspect of the same process, the primary mesenchyme cells being simply those cells that lose contact completely and are thus free to emigrate.

Primary invagination eventually stops. Then, after a short delay reflected by a plateau in the invagination-time curve, the second phase begins. This is always associated with the formation of fine pseudopia by the pulsating secondary mesenchyme cells, which are concentrated at the tip of the archenteron. These filopodia are often spun out directly from the pulsatory lobes. They extend into the blastocoelic fluid and eventually reach the inner aspect of the blastocoel wall, where they adhere and exert contractile tension (Fig. 10-3). It is the belief of both the Swedish and the Japanese schools that these filopodia exert sufficient tension to stretch and pull the archenteron to the animal pole to complete invagination.

In examining the evidence for this conclusion there are two main questions to ask: 1) do filopodia of the secondary mesenchyme exert

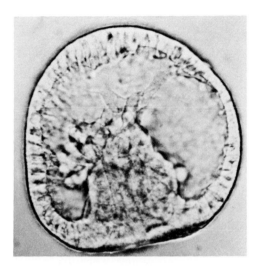

Fig. 10-3 Midgastrula of *Paracentrotus lividus*, showing filopodial extensions of secondary mesenchyme cells extending from the tip of the archenteron to the blastocoel wall near the animal pole. ×60. (Trinkaus. Unpublished.)

sufficient force to stretch the archenteron to more than twice its length, and 2) is the contractile pull they exert indispensible for the second part of invagination? A number of observations suggest that the answer to the first question is positive. Dan and Okazaki showed that the filopodia exert enough contractile tension to distort the form of objects to which they attach. The blastocoel wall is often pulled inward where filopodia attach, and if there are more filopodia attached to one side of the tip of the archenteron that side moves closer to the animal pole. If the blastocoel is caused to swell osmotically by sucrose, the filopodia are broken, and with this the tip of the archenteron retracts somewhat. Oil droplets injected into the blastocoel will attach to free secondary mesenchyme cells and be pulled to the animal pole as the cells contract. When gastrulae are caused to exogastrulate, no filopodial attachments are in evidence. Sometimes, however, such an exogastrula will begin to invaginate again; this only occurs if the archenteron is reattached to the blastocoel wall by filopodia. In these "entoexogastrulae" the force exerted may be so great that it stretches the archenteron wall almost to the breaking point. If it does rupture, it does so in the thinnest region, and the more animal part of the archenteron then goes on to the animal pole. Gustafson and Wolpert have made a rough calculation of the force required to bring about invagination on the basis of the force required to elongate a cylinder of appropriate dimensions and find it to be small, of the order of 10^{-2} dynes. From what we know of contractile forces exerted by the cytoplasmic fibers it is quite conceivable that several filopodia could collectively exert such a force. This would require only the tiniest fraction of the energy produced by a sea urchin egg, as measured by respiration, a ratio of available to required energy of about 10^6. This confirms the conclusions previously reached by Moore based on the amount of increased osmotic pressure created within the blastocoel that would just prevent invagination. From all this, it is evident that secondary mesenchyme filopodia collectively exert considerable contractile tension, which in all likelihood is sufficient to pull the archenteron to the animal pole.

Now to the second question, viz., is this contractile pull indispensible for the second phase of invagination? Gustafson and his colleagues have provided the main evidence here, taking full advantage of time-lapse cinemicrography (and indeed demonstrating in a most elegant way the utility of this method). They have followed the whole process of invagination in exquisite detail, tracing individual cells and making quantitative comparisons. The fundamental observation was that in all circumstances the second phase of invagination is correlated with attachment of the archenteron tip to the blastocoel wall by filopodia of the secondary mesenchyme. This is a dynamic process in which filopodia are being made, broken, and remade continually. Many of these filopodia are observed to shorten as the archenteron elongates. After completion of the primary phase, invagination stagnates somewhat and begins again or accelerates only when filopodial attachments are made. If secondary mesenchyme cells pull out of the tip of the archenteron, the rate of invagination slows down or stops and does not begin again until new filopodial attachments are made. As in the case of primary mesenchyme cells, these filopodia explore randomly and adhere to the blastocoel wall at the inner junctions of its cells (Fig. 10-4). Since the free primary mesenchyme cells clearly use these adhesions as points of attachment for traction as they migrate along the inner surface of the blastocoel wall (p. 33), it seems probable that secondary mesenchyme cells make similar use of them. There is an important difference, however. When a filopodium of a secondary mesenchyme cell contracts and its attachment remains firm, either its cell body will be pulled out of the tip of the archenteron, or tension will be exerted on the archenteron. In gastrulae vegetalized with lithium chloride, good secondary invagination is always associated with good filopodial attachments and poor invagination with poor attachments. Dan and Okazaki also observed that if filopodial activity is suppressed by chemical means (proteolytic enzymes, low calcium), secondary invagination fails. All of this evidence is very impressive. But we are reminded by Gustafson and Wolpert that correlations of this sort only support or disprove an hypothesis; they cannot constitute proof. In consequence, it cannot be said that the hypothesis is proved in a rigorous sense. We can say, however, that the support given by the accumulated evidence is most impressive, and we have every reason

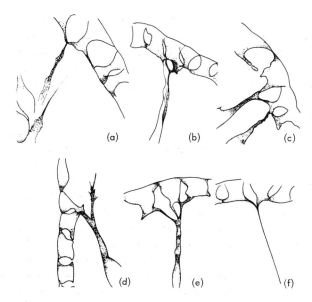

Fig. 10-4 Attachment of archenteron tip filopodia to the ectodermal blastocoel wall. (a) Simple attachment and slight "attachment cone" formation, (b) two filopodia attaching to the same site, (c and e) double and triple attachment of branched filopodia, (d) attached filopodia continuing its exploration with a branch, (f) attachment of a thin filopodium and the formation of an "attachment cone." (Gustafson. 1963. Exptl. Cell Res. **32**:570.)

to accept provisionally the hypothesis that the contractile pull of the secondary mesenchyme filopodia is both sufficient and indispensible for the second phase of invagination.

Nevertheless, there are questions that remain. In *Amphioxus,* where invagination takes place in about the same way as in the sea urchin (except that the invaginating region is larger), the process proceeds to completion in the absence of contracting filopodia. This suggests the possibility that other means could operate to cause elongation of the archenteron during the second phase of sea urchin gastrulation. Careful examination of the sea urchin data reveals that the correlation between the beginning of the second phase of invagination and the presence of connecting filopodia is not always so positive as it appears at first glance. Sometimes the archenteron appears to continue to invaginate after completion of the first phase but before filopodial attachments have been made. Could there be a small amount of autonomous extension of the archenteron at the beginning of the second phase? The attachments of filopodia to the blastocoel wall have been carefully traced, and, as expected, they tend to attach to the regions of the wall close to the tip of the archenteron. When invagination has proceeded more than halfway this will be predominantly the animal pole region and all is well. But at the completion of the first phase of invagination the archenteron tip is only one-third of the way and is in fact nearest the more vegetal areas of the blastocoel wall. Filopodia would accordingly be expected to attach initally preponderantly to these regions. Photographs presented by Gustafson and Wolpert appear to fulfill this expectation.

We ourselves have studied invagination in another sea urchin, *Lytechinus variegatus* (which is found in tropical waters of the western Atlantic) and have also observed numerous attachments to the vegetal areas as the second phase begins. In this situation, we would expect invagination of the archenteron during the first part of the second phase to be opposed by the contractile tension exerted by many or even most of the filopodia. Nevertheless, invagination proceeds. This suggests that some autonomous extension may occur at this time. A characteristic feature of the second phase of invagination is the thinning of the archenteron wall (exhibited by any epithelial sheet as it expands). Sometimes in *Lytechinus* this thinning begins before filopodial attachments to the animal pole region are evident. All this was too preliminary to allow us to do more than raise a tentative question. Is it possible that some autonomous extension of the archenteron wall is occurring and that secondary invagination is due both to it and to the external pull of the secondary mesenchyme filopodia? If so, the filopodia of the secondary mesenchyme would have a dual role. They would exert both a pulling force and at the same time anchor the archenteron in place, so that as it extends (and builds up much contractile tension within itself) it is not in danger of retracting.

Because of these apparently contradictory observations I reinvestigated this situation in gastrulae of *Paracentrotus lividus* (a Mediterranean sea urchin) and found that in this species filopodia attach predominantly to the animal region at all times, even at the onset of the secondary phase. This observation is consistent with the filopodial contraction hypothesis and supports it as a complete explanation of the secondary phase of echinoderm invagination. Since the filopodia of the secondary mesenchyme cells apparently adhere preferentially to the blastocoel wall in the general area of the animal pole, it seems likely that the inner surface of the blastocoel wall is more adhesive in the animal pole region. I also obtained evidence for the anchoring role of the filopodia. If the filopodia are broken by compressing the egg in a rotary compressor used for immobilizing Protozoa, the archenteron immediately retracts.

Movement by means of the contraction of adhering filopodia has emerged as a major means of cell locomotion utilized by such diverse cell types as sea urchin mesenchyme, dissociated sponge cells, chick retinal pigment cells, and deep cells of teleost blastoderms. It is therefore important that this type of movement be given much more attention than it has yet received. The glasslike transparency of a number of echinoderm eggs makes them excellent material for such investigations, and some progress has already been made in the study of the directed movements of the primary mesenchyme cells (see p. 33). A number of questions are begging answers. To what extent do differences in contact of filopodia with ectoderm cells really correspond to differences in adhesiveness of the latter, and to what degree are they due to other factors that may contribute to cell contact? It has been proposed that the filopodia of secondary mesenchyme cells move up the dorsal wall of the blastocoel because the cells of this region are more adhesive. Yet primary mesenchyme cells do not lodge there. Moreover, filopodia do not attach to the animal plate at the animal pole even though these cells seem to be more adhesive, as judged by the large amount of their lateral surfaces in apposition. What kinds of attachments are made to the blastocoel wall by the tips of the secondary mesenchyme filopodia? Also, what binds secondary mesenchyme cells to the tip of the archenteron, so that contraction of their filopodia can exert tension on the archenteron wall? These problems should be studied at both the optical and fine-structural level. How do the tips of the filopodia move over their substratum, the inner surface of the ectoderm? Is it proper to consider these filopodia as elongate ruffled membranes? Is there a way of measuring the tension in the filopodia? In studies of another contractile system, the tail ectoderm of the ascidian tadpole, Cloney has recently demonstrated the presence of oriented filaments in the apical cytoplasm of each cell. It seems probable that they lie at the basis of ectodermal contractility. Taylor has also found microtubules in small filopodia and cortical cytoplasm of cells in culture. Perhaps the electron microscope would reveal filaments or microtubules in the contracting filopodia of sea urchin mesenchyme.*

Wolpert and Gustafson are much impressed by the apparent similarity of several morphogenetic events in early echinoderm development and have made a provocative attempt to explain all in terms of two cell properties: cellular motility and differential cellular adhesiveness. Since

*(Note added in proof). Tilney, Gibbons, and Porter have just published such a study. Their micrographs show large numbers of microtubules distributed parallel to the long axis of the filopodia of primary mesenchyme cells. After the filopodia fuse to form a "cable" of cytoplasm (see Fig. 2-6), microtubules become aligned parallel to the long axis of the cable and parallel to the cytoplasmic stalks connecting the cell bodies to the cable. When colchicine and hydrostatic pressure (which cause microtubules to disassemble) are applied to gastrulae, the microtubules disappear and the primary mesenchyme cells tend to spherulate. When D_2O (which tends to stabilize microtubules) is applied, the microtubules persist and the cell asymmetries remain unaltered. From these results, it is concluded that microtubules are influential in the development of the filopodial form of primary mesenchyme cells.

the only evidence so far is descriptive (cell movements and shape changes as observed with time-lapse), these postulates at present lack proof. The arguments, however, are persuasive, and a careful examination of the reasoning will profit anyone interested in the mechanism of form changes in development (see Ch. 11). At the very least, the effort is refreshing in contrast to the usual pessimistic attitude that such processes are so complex as to almost resist analysis. Moreover, this hypothesis points up the pressing need to find independent objective means of studying the adhesiveness of different regions of the blastula and gastrula and mechanisms whereby the mesenchyme cells move.

SELECTED REFERENCES

Dan, K. and K. Okazaki. 1956. Cyto-embryological studies of sea urchins. III. Role of the secondary mesenchyme cells in the formation of the primitive gut in sea urchin larvae. Biol. Bull. **110**:29-42.

Gustafson, T. and H. Kinnander. 1956. Microaquaria for time-lapse cinematographic studies of morphogenesis in swimming larvae and observations on sea urchin gastrulation. Exptl. Cell Res. **11**:36-51.

Gustafson, T. and L. Wolpert. 1967. Cellular movement and contact in sea urchin morphogenesis. Biol. Rev. Cambridge Phil. Soc. **42**:442-498.

Hörstadius, S. 1939. The mechanics of sea urchin development, studied by operative methods. Biol. Rev. Cambridge Phil. Soc. **14**:132-179.

Moore, A. R. and A. S. Burt. 1939. On the locus and nature of the forces causing gastrulation in the embryos of *Dendraster excentricus*. J. Exptl. Zool. **82**:159-171.

Okazaki, K., T. Fukushi, and K. Dan. 1962. Cyto-embryological studies of sea urchins. IV. Correlation between the shape of the ectodermal cells and the arrangement of the primary mesenchyme cells in sea urchin larvae. Acta Embryol. Morphol. Exptl. **5**:17-31.

Tilney, L. G. and J. R. Gibbons. 1969. Microtubules in the formation and development of the primary mesenchyme in *Arbacia punctulata*. II. An experimental analysis of their role in the development and maintenance of cell shape. J. Cell Biol. **41**:227-250.

ELEVEN

Amphibian Gastrulation

Although the nature, extent, and timing of the movements of gastrulation in amphibian eggs have been described in detail by a number of workers, the analysis of these movements in modern times has rested largely in the hands of one man, Johannes Holtfreter, and were it not for his researches of the forties we would have little to say about the mechanism of these movements today. In these studies he dissected living gastrulae to observe cell relationships, he combined intact pieces of germ layers in culture, and he dissociated cells and studied their behavior in aggregates. He also observed the behavior in culture of individual cells from the different germ layers.

Holtfreter placed great emphasis on the distinctive characteristics of the surface of the egg and contended that all surface cells are held together by a pigmented extracellular surface gel layer to which they adhere, the famous "surface coat." This layer is still ill-defined as regards some important qualities, such as its composition and dimensions and whether it is really a supracellular structure to which cells adhere or instead simply a composite of the outer differentiated membranes of the tightly cohering surface cells. The electron micrographs of some workers reveal no unifying extra layer applied to the electron-dense boundaries of superficial cells. Others, however, find a layer of variable density and thickness applied to the surface of early blastomeres in some forms. It is not clear what these contradictory results mean, but it should be emphasized that if the coat consisted of mucopolysaccharides, as seems likely, it would have very low electron density and be difficult to visualize in the electron microscope, except with special staining techniques (see p. 62). What then are we to make of Holtfreter's concept of the surface coat? Even though the coat as an anatomical entity is in doubt, the observations on the behavior of the superficial cell layer remain valid. Cells within the layer cling tightly to each other and are thus united into a single cell layer, with a nonadhesive outer surface and an adhesive inner surface (as judged by adhesiveness to glass). The tight cohesion of these cells could be due to close adhesion of their lateral boundaries. Electron micrographs indicate that this may well be the correct explanation. Apparently the distal regions of the plasma membranes of these cells are closer to the boundaries of the next cell (70–90 A, as opposed to 110–200 A more proximally).

The outer cell layer is normally under considerable tangential contractile tension; this causes isolated pieces to curl with the outer surface on the inside. Inner cells of the egg, unattached to the surface layer, are apparently much less adhesive; they can be much more easily pulled apart with needles than surface cells. It appears that during all of gastrulation the majority of the original surface cells never become detached from each other, even after involution. thus, in a certain sense this cell layer never leaves the surface of the egg. It comes to line the archenteron wall, but the archenteron, like the intestine, is really a continuation of the outer surface of the organism. This fact in itself suggests that the tight adhesion of this layer plays a very important role in gastrulation.

Just prior to the beginning of invagination certain surface cells in the region of the prospective blastopore move inside and assume an elongate bottle shape. These are the so-called "bottle cells" or "flask cells." These prospective endoderm cells progressively attenuate until the bulk of each cell has moved completely into the interior. Significantly, however, they rarely lose their connection with the surface, even though the neck of the bottle may be stretched enormously (Fig. 11-1). It has recently been shown

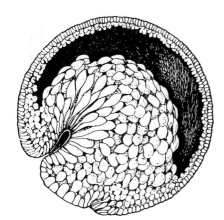

Fig. 11-1 Schematized section through an amphibian gastrula showing elongate bottle cells. (After Holtfreter. 1943. J. Exptl. Zool. **94**:261.)

148   *Amphibian Gastrulation*

in the electron microscope that bottle cells protrude long microvilli at their distal surfaces, which interdigitate with adjacent cells. They also have lateral interdigitations with adjacent cells. It may be that by these means bottle cells are anchored to the archenteron surface (Fig. 11-2). Because of their close association with the first indication and continued enlarge-

Fig. 11-2   The distal end of a neck or flask cell of an amphibian gastrula, showing lip (fl) of cell, interlocking of flask (fc) and wedge cells (wc) in region of dense layer (dl), and faint fibrils in dense layer. bg, blastoporal groove; v, vesicles in vesicular layer. ×65,000. (Courtesy of P. Baker. After Baker. 1965. J. Cell Biol. **24**:95.)

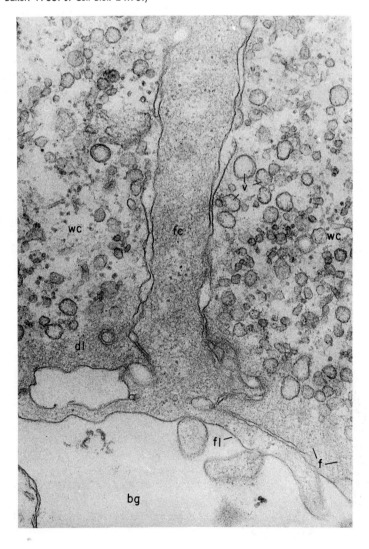

ment of the blastopore, these cells have attracted the attention of students of gastrulation for years. In fact, much of what I am about to describe was observed years ago by Rhumbler, Ruffini, Vogt, Waddington, and others. It is Holtfreter's analysis, however, which has led us closest to what appears to be a correct concept of their function. He showed by dissection of living eggs that wherever a cluster of these cells appears the surface is slightly indented and condensed. Adjacent nonbottle cells tend to stretch and orient radially around these centers of identation. As bottle cells become more and more attenuated and as their number increases, the slight indentations of the surface deepen and merge to form the beginning invagination of an early gastrula. They are now called "neck cells." The attenuation of these neck cells reaches its maximum when the archenteron has penetrated halfway into the blastocoel (see figure 11-2), but they are confined to the anterior front of the invaginating mass. The posterior and lateral regions of the archenteron, which we know from Vogt's marking studies have already moved over the lips of the blastopore in a movement of involution, are composed of an epithelium of short, roundish cells. Ruffini remarked that bottle cells are associated with invaginations of many kinds: optic vesicle, nasal placode, mouth opening, etc. There seems little question, therefore, that bottle cells of various sorts are important in invagination. The question is how. Do they play an active causal role, or are they merely a natural consequence of infolding?

Rhumbler proposed back in 1902 that the formation of bottle cells is the main cause of invagination, and he went on to suggest that the great expansion of the membranes of these cells on their internal aspect is due to a gradient in surface tension between the inner and outer milieu of the late blastula. A decrease in surface tension within the blastula was thought to transform cell shape into the flask form, with consequent invagination. This possibility has since received support from two Dutch workers, who showed that there is a pH gradient in amphibian embryos, the blastocoel fluid being more alkaline (pH 9) than the cells (pH 6.7 to 7.6). If these determinations can be relied upon they are of great interest, because alkalinity has a profound effect on the cell surface, causing formation of pseudopodia and greater cell motility. On this basis Holtfreter, following Rhumbler, proposed that the lower surface tension of the blastocoel fluid causes the inner surface of endoderm cells to expand and move into the interior. He emphasizes, however, that the connection of these cells to the surface is essential for invagination. Otherwise they would slip into the interior as individuals.

This was substantiated by an elegant experiment. A group of surface endoderm cells from the blastopore region was isolated and placed on a substratum of inner, blastocoelic endoderm in culture. Under these conditions the graft immediately adheres to its substratum and within 1 to 2 hours sinks into it. Dissections reveal that during this process the contracted neck cells extend and once again form neck cells, which squeeze between the cells of the substratum. Most of them remain attached to the surface layer, which, due to its low adhesiveness, is not engulfed by the surrounding cells. The pulling force of the deeply rooted cells is so great, however, that the cell layer at the surface is dragged partly into the mass of the substratum, and a beginning blastopore is established (Fig. 11-3). Thus, beginning invagination is beautifully simulated. If this experiment is repeated with deep endodermal cells, they too invade the endoderm substratum and become stretched more or less perpendicular to the substratum. But they do not form neck cells; they scatter more freely among the host cells, and no blastoporal groove is formed.

These experiments demonstrate that isolated endodermal cells of the blastopore can penetrate into a proper cellular substratum and assume a bottle shape. But they do not drag along in their wake a tapering neck or form an invagination unless they remain attached to the surface. It appears therefore that they and the suface layer to which they remain attached are responsible for the formation and deepening of the blastopore. It is of interest in this connection that the initiation of ventral invagination during normal gastrulation is also characterized by bottle and neck cells. In this region, however, the neck often breaks, and the cells move into the interior as individuals. Perhaps this accounts in part for the limited

Fig. 11-3 A graft of amphibian blastoporal cells sinking into an endodermal substratum (left to right) to form a blastoporal groove. Upper row, sections; lower row, surface views. (After Holtfreter. 1944. J. Exptl. Zool. **95**:171.)

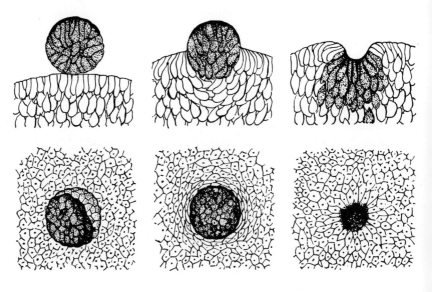

extent of this invagination and is nature's experiment in support of the postulated role of the bottle cells. The superficial similarity of primary invagination in amphibians and sea urchins is striking. In both cases invagination depends on the centripetal activities of certain cells of the endoderm and the presence of a lowly adhesive surface layer.

In the light of the probable role played by the contracting necks of the bottle cells, it is significant that the long necks of these cells have been shown to possess microfilaments in their distal cytoplasm (Fig. 11-2). These filaments could have a contractile function (but probably tangentially, not proximo-distally).

Invagination in endodermal isolates takes place in Holtfreter's solution, hence probably in the absence of the postulated pH gradient within the egg. This suggests that the formation of bottle cells is not dependent on such gradient. Holtfreter assumes that "the inner milieu considerably lowers the surface tension of the proximal side" of the cells in such explants and causes them to elongate. But since there are no measurements of pH or surface tension within such explants the assumption is without basis. Incidentally, the Rhumbler-Holtfreter hypothesis does not explain why only endoderm cells of the marginal zone (in the region of the prospective blastopore) form bottle cells. Surely others, in the yolk plug, are equally exposed to the postulated pH gradient. The coup de grace to the hypothesis was administered by Stableford, who showed that when he opened the blastocoel of early gastrulae, thus eliminating a pH gradient, and placed eggs in media of various pHs, invagination began and continued willy-nilly.

All this leaves us with the following conclusion. Bottle cells and the surface layer with which they cohere are responsible for beginning and continuing invagination in amphibian eggs. The tightly cohering surface cell layer functions not only in integrating the pulling activity of independently moving bottle cells into a coordinate system; it also communicates the pull to other endoderm and mesoderm cells. Moreover, because of its nonadhesive surface it probably prevents mutual adhesion of the marginal zone and the yolk plug at those points where they contact each other, as both slip into the interior. Since bottle cells arise independently of the medium, their formation appears to be due to their association with their cellular substratum. If we reasoned in the fashion of Steinberg (Ch. 7), we would expect the blastoporal cells to possess greater adhesiveness than the inner endodermal cells. But then why would they be in the surface layer? Their adhesion to the very lowly adhesive surface layer could explain this.

Once invagination is begun by the bottle cells, what forces cause it to continue? One old theory suggests that the continuing mesodermal invagination is the result of pushing by the expanding ectoderm. This pushing force, however, if important at all, is completely dispensable, for when the entire ectoderm is removed, invagination and internal stretching continue more or less as usual. The cause seems to lie in the marginal zone itself. Schechtman has shown that if a graft of dorsal lip is placed on inner blastomeric endoderm, both the individual cells of the graft and the graft itself elongate along their anteroposterior axes. Apparently, the cells of the marginal zone are endowed with an autonomous capacity to stretch in the proper direction. This substantiates the well-known tendency for dorsal lip to invaginate when grafted ectopically (Spemann and H. Mangold, Vogt, Lehmann) and could be a dominant factor in its continuing invagination. If deep mesoderm is grafted to deep endoderm, it moves into it and is engulfed. If the surface mesoderm is used, the cells move in but are not completely engulfed. The result is a short archenteron, demonstrating once again the importance of the tight cohesion of the surface layer for invagination. In normal gastrulation, of course, the invaginating mesoderm does not penetrate the endoderm; it glides into the archenteron on the surface of the deep endoderm and in this manner contributes to the enlargement of the archenteron. It is somehow prevented from invading the endoderm. The following experiment suggests that it is impeded by adhesion to the ectoderm. When the ectoderm of a neurula is removed, the mesoderm contracts into patches and sinks into the endoderm. The orderly, integrated involution of mesoderm is thus seemingly due to a number of factors. Among these are the autonomous tendency for mesodermal cells to stretch in an anteroposterior direction and to invaginate, restraint and coordination by its surface layer, and adhesion to the overlying ectoderm.

But does this suffice to explain the continuing movement of the mesoderm over the lips of the blastopore and then anteriorward as part of the wall of the archenteron, after invagination is well underway? An impressive feature of invagination is that once invaginating cells have moved into a mass of other cells they will move in again, if under experimental conditions they are presented with a second chance. Hence, the same forces that get them there keep them there, a conclusion of possibly far-reaching significance. But what are these forces? Perhaps a clue as to what is going on here is provided by the fact that no one has yet observed invagination to take place in the absence of an embedding matrix of other cells. Clearly the activity of the cellular substratum is of great importance. This is no surprise. Invasion is always accompanied by engulfment; the invaded tissue actively spreads over the invader. In invagination this is the ectoderm, whose spreading activity in epiboly causes it finally to engulf entirely both the mesoderm and the endoderm. Incidentally, there are many examples of cell masses consisting of hundreds of cells gliding into a cellular matrix with an apparently smooth periphery. This has been observed, for example, for neural plate, prospective forebrain, and a lump of mesoderm. Of course, it is not certain that the surface is really "smooth." Closer examination with modern methods will probably reveal cellular activity. But the point is that the active partner in such a system may not be the invader. It could well be the invaded tissue which spreads actively over its "invader."

Let us then move back to the surface of the egg, where the ectoderm is undergoing extensive epibolic spreading. Epiboly is aided by the migration of cells to the surface from lower layers, but it is due primarily to the spreading of cells already at the surface at the onset of gastrulation. What forces are responsible for this spreading? Is the ectodermal sheet pulled and stretched by the mesoderm as it rolls over the blastoporal lips into the interior? Or does the ectoderm possess spreading properties of its own? Holtfreter has also directed his attention to this problem. An isolated piece of surface ectoderm from any part of a gastrula will spread over any nonsurface tissue of the same age, such as endoderm or mesoderm. An autonomous capacity to spread is therefore an inherent property of any part of the ectoderm. But it will not spread unless attached to the substratum. Moreover, only the surface layer of the ectoderm will spread. The inner layer of ectoderm behaves entirely differently. It will move underneath surface ectoderm and will invade endoderm. The autonomous spreading of pieces of ectoderm suggests that individual ectodermal cells may possess the capacity to spread independently. To check this, Holtfreter dissociated ectoderm into individual cells and cultured them on glass. The cells adhere to each other and to the glass almost immediately and then, after about 30 min, flatten and spread. Apparently, ectoderm cells spread more frequently than isolated mesoderm or neural cells. The spreading of an ectodermal sheet appears, therefore, to depend on the spreading activities of its constituent cells. Otherwise, indeed, how could such a sheet spread without possessing a free edge? Exactly how the spreading of individual cells occurs within the sheet is not understood. The regimentation of individual cells into the coordinate spreading of the ectodermal sheet is no doubt dependent upon their tight cohesion. Yet, as the sheet spreads it is put under much tension, as demonstrated by the fact that a cut in it widens immediately. This stretched condition is no doubt due in part to its continuity with the invaginating prospective chordamesoderm. It would seem likely, therefore, that individual cells would pull apart from time to time, in spite of their overall tight cohesion, and then close the opening by active spreading, as in a contact-inhibiting system. This should be looked for.

In view of these studies, the strikingly different behavior of superficial and inner ectoderm deserves further study. It would be interesting to know, for instance, whether dissociated deep ectoderm cells behave like superficial ectoderm cells, both in cell culture and in mixed aggregates. Superficial ectoderm cells move to the surface of an aggregate. Would inner ectoderm cells do the same?

One final question concerning the ectoderm arises from the observation that an isolated piece of ectoderm will expand centrifugally in all directions. It possesses no inherent polarization. How then account for the oriented spreading of the ectoderm toward the blastopore in epiboly? It is not necessary to postulate contact guidance by an oriented substratum. By replacing the mesoderm as it undergoes involution, the spreading ectoderm simply follows the path of least resistance and through no effort of its own is bound to spread in an oriented fashion. The extent of the spreading of the ectoderm seems to be determined by the ready availability of the surfaces of deep cells. This is a nice illustration of how the oriented movement of one layer (in this instance the mesoderm) can impose orientation on another, so that its movement is coordinated with the pattern of the whole gastrula.

Endoderm is also a spreading layer, and much of what we have said about the ectoderm can be repeated here. During the course of gastrulation the endoderm spreads extensively inside the archenteron, in a dorsad movement of the walls of the archenteron, and later in increasing the surface area of the elongating gut. Superficial endoderm isolated from the archenteron will spread over deep endoderm. If the ectodermal surface layer of a neurula is removed, the endoderm lining the archenteron will move out through the blastopore and spread over the mesoderm to cover it. Thus, under appropriate conditions the endoderm may engage in reverse epiboly. It does not spread out of the blastopore during normal gastrulation, because it is opposed by the spreading ectoderm. In a sense, therefore, the two layers are in competition. But it can be shown that the ectoderm is the stronger competitor. If the ectoderm is partly removed, the endoderm will spread out of the blastopore. However, it will soon be stopped by the ectoderm, which has spread even more extensively (Fig. 11-4). As the two layers meet, the nonadhesiveness of their surfaces probably prevents mutual overgrowth. Here again the integration of the spreading of a layer with the pattern of the whole egg depends on its relation to another layer. The extensive spreading capacity of the endoderm is limited to an amount proper for coordinated gastrulation by the spreading ectodermal sheet. The extensive epibolic spreading of the ectoderm is therefore necessary to maintain the archenteron. The expansion of the coated endoderm sheet seems, like the ectoderm, to rest on the spreading activities of its constituent cells, as shown by their tendency to flatten in culture when dissociated. From all this, it would appear that we have some understanding of the spreading of the endoderm. There are, however, some important unknowns. Before spreading inside the archenteron the endoderm must get

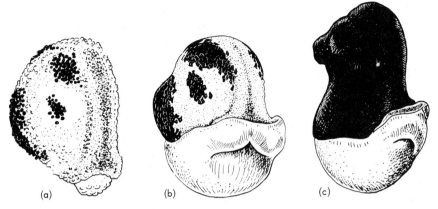

Fig. 11-4 *Rana pipiens* neurula, partially denuded of epidermal cells by alkali treatment. Remaining epidermis and gut epithelium compete in spreading over the subepidermal tissues. (Holtfreter. 1944. J. Exptl. Zool. **95**:171.)

there. Why is there no spreading of the endoderm prior to invagination? More than this, what permits the endoderm to contract as invagination begins and form bottle cells? These are striking changes in endoderm behavior for which we have no adequate explanations.

In his analysis of amphibian gastrulation, Holtfreter placed much emphasis on differences in cellular adhesiveness and their importance in determining whether a layer will infiltrate its cellular substratum or spread on its surface. In this he anticipated the later postulates of Steinberg. He showed that gastrula mesoderm is engulfed by endoderm and concluded that this indicates greater cohesiveness of the mesoderm. Deep endoderm is in turn engulfed by superficial ectoderm. If differences in adhesiveness are indeed crucial, superficial ectoderm would be the least cohesive of all, and one might predict that a chunk of mesoderm would be engulfed by superficial ectoderm, which of course it is. So we have a hierarchy of cohesiveness from the most to the least: mesoderm, endoderm, ectoderm. But how then do we explain the behavior of these layers during normal gastrulation and their disposition at the end? Mesoderm, which should be in the center of the gastrula because of its greater cohesiveness, is of course between the other two. The modifying effect of the nonadhesive ectoderm surface could be critical. As we have noted, inner ectoderm behaves very differently and is engulfed by endoderm, rather than the reverse. It would be of great interest to know if inner ectoderm would likewise be engulfed by mesoderm. Fortunately, this experiment has just been performed. Inner ectoderm is indeed engulfed by mesoderm. With this result we can solve the problem by reasoning that ectoderm cells are actually the most cohesive of all but have their cohesiveness drastically reduced by their nonadhesive surface. Steinberg has emphasized that in such a situation the ectoderm would be at the periphery, with its nonadhesive outer surface turned to the outside, because its average cohesiveness is the lowest. But since the inner surface of the ectoderm is the most cohesive, mesoderm, which is more cohesive than endoderm, would take the intermediary position and adhere to the inner surface of the ectoderm.

Incidentally, since the mesoderm is less cohesive than the inner surface of the ectoderm, it would tend to spread on it. This could account for the expansion of the roof of the archenteron and in part for mesodermal invagination. Later on, endoderm would spread on the inner surface of the more cohesive invaginated mesoderm, and the proper disposition of the germ layers would be achieved. Everything depends on the peculiar properties of the ectoderm. If the ectoderm is removed we would expect the mesoderm to sink into the endoderm. It does just this. In consequence, by simply assuming regional quantitative differences in adhesiveness we can explain how complex tissue organization can arise. The extreme importance of the nonadhesive outer ectoderm layer is emphasized. Indeed, it prevents the embryo from turning inside out. Remove this ectoderm layer, and this is precisely the result—the endoderm now has a free edge and spreads out and over the egg (Fig. 11-4).

These postulations accentuate the need for quantitative studies of cellular adhesiveness. A recent study by Johnson represents a first step in this direction. He compared the contact behavior of dissociated presumptive mesodermal cells from normal amphibian embryos and interspecific hybrid embryos whose development is arrested at gastrulation. Normal cells have a strong tendency to flatten on a glass substratum and reaggregate. Cells from three different arrested hybrid gastrulae, in contrast, have a markedly lower tendency to do so. This constitutes strong evidence that a certain degree of cellular adhesiveness is necessary for gastrulation.

Two other studies suggest that cellular adhesiveness increases during the course of amphibian gastrulation and that calcium is involved in this increase. As the relatively loosely cohering cells of the blastula become arranged into tightly cohering germ layers, the cells become more difficult to dissociate. Jones and Elsdale found that cells of the blastula will dissociate in a calcium-free medium, whereas those of the neurula require EDTA and those of post-neurula stages require EDTA plus proteolytic enzymes. In view of the oft postulated role of calcium in cell adhesion, Stableford determined the calcium concentration of embryos during blastula and neurula stages. His findings are of considerable interest. During this period of increasing cellular adhesiveness, the calcium content of the blastocoel fluid increases eightfold, to a level almost as high as that in Holtfreter's solution. It is significant that Holtfreter's solution causes blastula cells to cohere more tightly than they do normally.

In summary, we now possess the following picture of the mechanism of amphibian gastrulation. Cells become more adhesive and gastrulation begins as the endodermal cells of the blastoporal region sink into the deep endoderm to form bottle cells. Because their outer ends are tightly coherent, the surface is indented and other endoderm cells are dragged in; thus the archenteron is formed. The prospective mesoderm follows, apparently in part because of an inherent tendency to stretch in an anteroposterior direction. It then spreads along the more adhesive undersurface of the ectoderm to enlarge and complete the formation of the archenteron. Meanwhile, the ectoderm cells express their inherent spreading capacity by expanding over the animal hemisphere to engulf the invaginating mesoderm and replace it as it disappears from the surface. The endoderm possesses properties similar to the ectoderm and in consequence spreads extensively to line the archenteron and aid in its enlargement. The greater spreading tendency of the ectoderm confines the spreading of the endoderm to the lining of the archenteron and thus prevents a reversal of involution.

This is rather a neat picture, with an inner consistency and reasonably convincing documentation. It therefore serves well as a basis for continuing investigation. We need to know more about the ectoderm. If it did not play the role assigned to it the whole structure would collapse. What are the fine-structure relations of cells of the various germ layers, particularly between invaginating cells and their invagination substratum? And then there is the old and difficult problem of cell surface adhesiveness. Do the postulated quantitative differences actually exist? How can we account for regional differences in the behavior of marginal-zone and yolk-plug endoderm cells, with some forming bottle cells and others not? By what mechanism do cells adhere to each other during these extensive movements? Are there specializations of the cell surface, like close junctions and desmosomes, from the beginning of movement; or do they only appear later? When do ectoderm and endoderm cells acquire spreading capacity, and how does spreading occur in a sheet that lacks a free edge? Finally, what factors cause the various changes in cell shape? Are microtubules or microfilaments involved, or can these changes be attributed to modifications of cell surface properties?

Waddington made a rather ingenious attempt to measure the force produced by the embryo during amphibian invagination by opposing it with an iron ball placed in the gastrula attracted by an electromagnet. He then measured the magnetic field required just to prevent invagination. He arrived at values of approximately 3.4×10^{-5} dyne, a force considerably smaller than appears necessary for sea urchin invagination (see p. 140). Of course, one has no way of knowing in such an experiment whether the opposing force has altered the magnitude of the morphogenetic force.

SELECTED REFERENCES

Baker, Patricia A. 1965. Fine structure and morphogenetic movements in the gastrula of the tree frog, *Hyla regilla*. J. Cell Biol. **24**:95–116.

Balinsky, B. I. 1961. Ultrastructural mechanisms of gastrulation and neurulation. Symp. on Germ Cells and Develop., Inst. Intern. Embryol. Fondaz. A. Baselli. p. 550–563.

Holtfreter, J. 1943. Properties and functions of the surface coat in amphibian embryos. J. Exptl. Zool. **93**:251–323.

Holtfreter, J. 1943. A study of the mechanics of gastrulation: Part I. J. Exptl. Zool. **94**:261–318.

Holtfreter, J. 1944. A study of the mechanics of gastrulation: Part II. J. Exptl. Zool. **95**:171–212.

Holtfreter, J. 1947. Observations on the migration, aggregation, and phagocytosis of embryonic cells. J. Morphol. **80**:25–55.

Holtfreter, J. and V. Hamburger. 1955. Embryogenesis: progressive differentiation. Amphibians, p. 230–296. *In* B. H. Willier, P. Weiss, and V. Hamburger [ed.] Analysis of development. Saunders, Philadelphia.

Johnson, K. E. 1969. Altered contact behavior of presumptive mesodermal cells from hybrid amphibian embryos arrested at gastrulation. J. Exptl. Zool. **170:** in press.

Jones, K. W. and T. R. Elsdale. 1963. The culture of small aggregates of amphibian embryonic cells *in vitro*. J. Embryol. Exptl. Morphol. **11**:135–154.

Perry, M. M. and C. H. Waddington. 1966. Ultrastructure of the blastopore cells in the newt. J. Embryol. Exptl. Morphol. **15**:317–330.

Rhumbler, L. 1902. Zur Mechanik des Gastrulationsvorgänges, insbesondere der Invagination. Eine entwicklungsmechanische Studie. Arch. Entwicklungsmech. Organ. **14**:401–476.

Schechtman, A. M. 1942. The mechanism of amphibian gastrulation. I. Gastrulation promoting interactions between various regions of an anuran egg (*Hyla regilla*). Univ. Calif. Publ. Zool. **51**:1–40.

Stableford, L. T. 1949. The blastocoel fluid in amphibian gastrulation. J. Exptl. Zool. **112**:529–546.

Steinberg, M. S. 1964. The problem of adhesive selectivity in cellular interactions, p. 321–366. *In* M. Locke [ed.] Cellular membranes in development. Academic Press, New York.

TWELVE

Neurulation

The formation of the rudiment of the central nervous system of vertebrates is in a sense an even more spectacular process than gastrulation. The elaboration of the neural plate and out of it the neural tube has fascinated embryologists for a long time, because it is the most obvious result of primary induction by the chordamesodermal roof of the archenteron. It is indeed the best known induction of a morphogenetic movement in a cell sheet by cells adjacent to the sheet. In amphibians (where it has been studied the most), reptiles, birds, and mammals, the neural tube forms largely out of the surface ectoderm in a manner that in its gross aspects is easy for all to see (Fig. 12-1). It involves a large proportion of the surface of the egg, including as much as 50% of the ectoderm of an amphibian egg. It occurs relatively rapidly and can be readily followed with vital-dye marking and time-lapse cinemicrography. In point of fact, the descriptive aspects of the folding movements of the neural plate during formation of the neural tube and the prospective fate of each part of the neural plate during development have been studied so exhaustively with vital dyes and transplantation that from a descriptive point of view neurulation is one of the best understood aspects of development. It is therefore particularly frustrating that the mechanism whereby this deceptively simple process occurs has until now resisted all analysis. Valiant attempts to discover the key have been made several times during the past century, beginning with Wilhelm His in 1874. But all proposals have either been since disproved or rest unproved for lack of sufficient evidence. It is nevertheless worthwhile to survey the various proposals. Not only is the mechanism of neurulation an important process in itself, but, involving as it does in some cases the

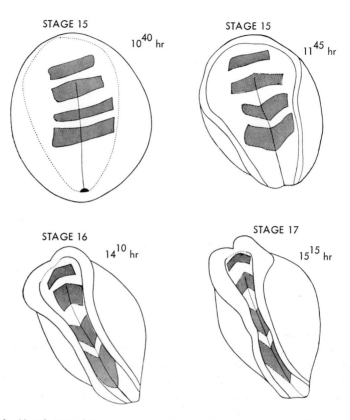

Fig. 12-1 Neurulation in the mexican axolotl. Note the changes in size and position of the color marks applied to the neural plate (shaded areas). (Jacobson. 1962. Zool. Bidrag. Uppsala. 35:433.)

folding of a sheet one cell thick, understanding of it will no doubt teach us much about the foldings of many other sheets of cells which are crucial for morphogenesis.

The neural tube forms in a similar fashion in all vertebrates except fish, where the neural plate sinks into the underlying tissues without forming a tube. (The tube forms later by some sort of hollowing-out process.) In the amphibia, the neural plate forms on the dorsal side of the embryo and quickly thickens due to an elongation ("palisading") of its cells in the radial plane of the embryo. It is perfectly symmetrical bilaterally, lying on either side of the midline between the closed blastopore and a point roughly above the anterior end of the prechordal mesoderm. The neural folds form at the edge of the plate and are accompanied by a further elongation of the neural plate cells in this region (Fig. 12-1), such that the plate thickens at this point and cells just lateral to the plate push up into the folds. These neural folds at the lateral edges of the plate now roll toward the midline, i.e., toward each other. When the folds meet at the midline, the plate is folded into a tube, with its former exterior surface now the inner surface of the tube. Thus the neural tube is formed. The neural crest forms where the folds meet. The ectoderm lateral to the neural plate is also carried to the midline, where it now forms the epidermis covering the neural tube.

During this sequence of events cells in the middle of the neural plate undergo a gradual elongation and narrowing of their apical ends (Fig. 12-2), proceeding from cuboidal epithelium of the beginning neural plate, to cuboidal wedge-shaped cells at the neural groove stage, and then to very elongate flask-shaped cells at final closure of the folds when the neural tube is formed. This packing of neural plate cells to adhere over more of their surfaces and thus cover collectively less of the surface of the egg has an inevitable stretching effect on the rest of the ectoderm. The ectoderm thins correspondingly, its cells changing from slightly columnar to flat in shape. As the neural plate cells become columnar their nuclei take up position in the basalmost part of the cells, toward the chordamesoderm, and elongate. Nuclei of nonneural ectoderm retain their more or less central position in the cell and remain spherical throughout neurulation.

It has been agreed since the beginning of research on this problem that the formation and rolling up of the neural plate must in some way be related to the orderly changes that take place in the form of its constituent

Fig. 12-2 Diagram illustrating the correlation between the formation of wedge-shaped cells and bottle cells of the medullary plate and neural tube formation in an amphibian embryo. Stippling indicates the location of microfilaments. (Baker and Schroeder. 1967. Develop. Biol. 15:432.)

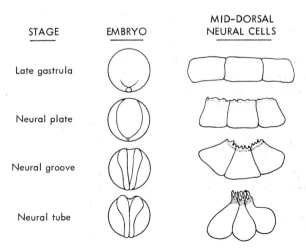

cells. The problem is whether these changes are the cause or the consequence of neural tube formation.

The earliest serious proposal was that of His in 1874. He suggested that the form changes in the cells and eventually in the neural plate as a whole result from growth pressures due to excessive mitosis in the neural ectoderm. If the plate were prevented from spreading by the surrounding ectoderm, then the increased cell number (with of course no decrease in cell size) would result both in a crowding of cells, hence a change to a columnar shape, and a buckling of the cell sheet, which in the right direction could cause a tube to form. This hypothesis waited over a half century to be tested. Gillette measured cross-sectional areas of both neural and epidermal ectoderm and counted the nuclei therein. By this method he established that there is indeed a 23% increase in the number of cells present in the whole ectoderm during neurulation, but he also found that this increase occurs without increase in total volume and that therefore cell size decreases. Mitosis is nonaccretionary. Moreover, and what is more important, no differences in change in cell size or cell number were detected between the neural and nonneural ectoderm. These results cast great doubt on the possible validity of the His hypothesis but fall short of a completely critical test. The areas examined were too large to detect possible local differences between, for example, the neural folds and the neural plate in between.

Another early hypothesis of neurulation was that of Glaser, who was impressed by the change in cell shape from a prismatic to a truncated pyramidal form. This hypothesis proposes that this change in cell form, which is a geometrical necessity for a folding epithelium, is responsible for the folding of the neural plate and that it is due to differentially higher water uptake by the more basal parts of the cells. This proposal was based originally on the belief that a localized increase in cell volume takes place in neurogenesis. This is clearly not the case, at least in *Amblystoma maculatum*, the form studied by Gillette. It has also been shown by density measurements that there is very little change in cell density during neurulation. But others have inhibited folding of the neural plate with hypertonic sugar solutions and caused rolled-up plates to unroll when treated with solutions of glycerine. The evidence is therefore contradictory. It also should be pointed out, however, that none of these tests examined possible differences in water uptake between the basal and apical parts of the cells, which is the really crucial part of the hypothesis.

A concept which has apparently received relatively little attention in the literature since first being proposed years ago by Rhumbler treats changes in curvature in terms of changes in the surface properties of individual cells. Although proposals of this sort are common currency in the present era of the sorting out of cells in aggregates, their application to the curvature of membranes has been surprisingly limited. Recently, such an approach has been promoted effectively by Gustafson and Wolpert for the gastrulation of echinoderm eggs. As may be recalled, this mechanism is based on the concept that the degree of contact between cells is determined by the adhesive force between them and the resistance of the cells to deformation. Applied to neurulation, one can easily see how increased surface adhesiveness coupled with resistance of the basal part of the cell to deformation could cause cells of the neural plate to assume an elongate wedge shape. This is certainly a possibility that deserves careful attention. But until now it has received none. Obvious first attempts to gain information on possible surface changes would be to study the relative ease of dissociating cells from different regions of the neural plate and epidermal ectoderm at different times of development, compare their contact and locomotory behavior in culture, and study their junctional contacts. Waddington and Perry and Baker and Schroeder have begun ultrastructural studies, and the first results reveal that neural fold cells are united by interlocking flanges of the cell surface and by what appear to be tight junctions and desmosomes. What is now required is a systematic study of various regions at various times.

A recurrent idea of a mechanism for causing the neural plate to roll up is by differential contraction of one surface of the sheet. Warren H. Lewis favored such a general mechanism for many morphogenetic movements and has been its primary proponent for explaining neurulation. There is as yet little evidence in its support, but the idea is by no means unreasonable; many multicellular systems have remarkable contractile properties. Very recently Baker and Schroeder have found bundles of microfilaments approximately 50 A in diameter in cells of the neural groove of *Hyla* and *Xenopus*. These microfilaments are located in apical cytoplasm just beneath the surface protrusions and are oriented parallel to the cell surface. It is proposed that these filaments are contractile and at least in part responsible for the narrowing of the apical ends of the cells as neurulation proceeds. The apical location and the orientation of the filaments are the primary evidence. And it is also pointed out that whereas the apical cell membrane becomes highly folded as the cells become more and more flask shaped, the filaments simply shorten and remain more or less straight.

This is a pretty correlation but not sufficient evidence for the hypothesis. There are many examples of cell membranes becoming highly folded and forming abundant microvilli in the complete absence of any narrowing or "contraction" of the subsurface cytoplasm. Moreover, Waddington and Perry, working with another amphibian egg, *Triturus alpestris,* found no such apical filaments in the necks of the flask-shaped cells during neurulation. Instead they found abundant microtubules in the cytoplasm oriented primarily in the direction of the cell elongation. They speculate that the main agent causing the shape change in these cells does not operate by lateral contraction but by radical elongation in which the microtubules play an important role. They also find microtubules with similar orientation in bottle cells of gastrulae. The evidence is obviously totally insufficient for both the "contracting" activity of the microfilaments and the "elongating" activity of the microtubules. But they are both interesting suggestions which may turn out to have validity. The use of treatments that impede microtubule and microfilament formation is an obvious next step in the analysis of this problem (see p. 207). We shall return to the possible importance of the cytoskeleton in cellular shape changes in the last chapter.

Holtfreter has contributed evidence that supports the possibility that internal factors may play a crucial role in the maintenance of the proximodistal polarity of these elongate cells. When cells of the neural plate are dissociated from each other and cultured as individuals they tend to stretch themselves along their original proximodistal axis into elongate cylindrical cells.

An elegant attempt has been made by Selman, in Waddington's laboratory, to measure the forces producing neural closure. He placed a tiny pair of dumbbell-shaped electromagnetically induced magnets on the neural plate, one against each neural fold. They were arranged so that they would repel each other when an electromagnetic field was switched on. The separation of the dumbbells was held constant so that the neural folds were held apart in the region of the dumbbells. The magnetic force which had to be applied in order to accomplish this was measured. Neurulae of *Triturus alpestris* exerted forces up to 4.0×10^{-2} and 4.5×10^{-2} dyne, and neurulae of the mexican axolotyl forces between 9.0×10^{-2} and 11.0×10^{-2} dyne. The magnitude of the energy requirements were therefore of the order of 10^{-2} erg for closure of the neural folds. According to Selman's calculations, these results show that neurulae can work against an external system at a rate which is 6 million times less than the total energy used by the embryo. These values confirm the conclusion reached earlier for amphibian and sea urchin invagination that morphogenetic movements in themselves require very little expenditure of energy. This statement is all the more true if one takes account of the fact that these measurements are of forces developed by embryos against an outside system, not forces that an embryo normally exerts against its own tissues. The estimates are therefore no doubt in excess of work performed naturally.

Having surveyed the main hypotheses for the mechanism of formation of the neural tube, we see that none of them affords us more than an entry into some of the possible ways in which neurulation may take place.

Considering the importance of the problem and the beauty of the material, it is striking that so little progress has been made. One of the reasons may be that hardly anyone has worked consistently on neurulation after his initial effort. This is not surprising for the past, when our knowledge of cell contact relations and cell movements was a good deal more primitive than it now is. But now, with many techniques and ideas available from the study of other systems, the time is surely ripe for a concerted attack on the mechanism of neurulation. It is to be hoped that as the research progresses we do not become overly wedded to one hypothesis or another. It seems quite possible, for example, that all three of the currently most popular forces proposed for changing the shapes of cells—changes in surface properties such as adhesiveness, contraction of oriented microfilaments, and elongation of oriented microtubules—are at work during formation of the neural tube.

SELECTED REFERENCES

Baker, Patricia C. and T. E. Schroeder. 1967. Cytoplasmic filaments and morphogenetic movements in the amphibian neural tube. Develop. Biol. **15**:432–450.

Brown, M. G., Hamburger, V. and F. O. Schmitt. 1941. Density studies on amphibian embryos with special reference to the mechanism of organizer action. J. Exptl. Zool. **88**:353–372.

Gillette, Roy. 1944. Cell number and cell size in the ectoderm during neurulation (*Amblystoma maculatum*). J. Exptl. Zool. **96**:201–222.

Holtfreter, J. 1948. Significance of the cell membrane in embryonic processes. Ann. N.Y. Acad. Sci. **49**:709–760.

Holtfreter, J. and V. Hamburger. 1955. Embryogenesis: progressive differentiation. Amphibians, p. 230–296. *In* B. H. Willier, P. Weiss, and V. Hamburger [ed.] Analysis of development. Saunders, Philadelphia.

Jacobson, C. 1962. Cell migration in the neural plate and the process of neurulation in the axolotl larva. Zool. Bidr. Upps. **35**:433–449.

Lewis, W. H. 1947. Mechanics of invagination. Anat. Record. **97**:139–156.

Selman, G. G. 1958. The forces producing neural closure in amphibia. J. Embryol. Exptl. Morphol. **6**:448–465.

THIRTEEN

Early Chick Morphogenetic Movements

The study of the mechanisms of morphogenetic movements during early chick development has dealt mainly with three distinct processes: epiboly of the area opaca, formation of the germ layers, and the migration of the precardiac mesoderm.

Epiboly of the area opaca

It has been known for some time from the work of Waddington and Spratt that chick blastoderms cultured on an agar or plasma-clot substratum do not expand. These conditions in culture interfered with their development and suggested that the normal substratum should be used. In ovo the upper surface of the marginal region of the blastoderm adheres to the under surface of the vitelline membrane and spreads on it. On the basis of this, New tried using the vitelline membrane as the substratum in culture and found that blastoderms spread on it in vitro as well as in vivo.

Here too, as in the spreading of other cell sheets, the nature of the substratum is crucial. With this in mind Bellairs examined the fine structure and composition of the vitelline membrane and found it to be composed of two protein layers, which differ both structurally and chemically. This is of interest, because although the blastoderm margin will adhere to both the upper and lower surface of the vitelline membrane it will spread only on the lower. This observation suggests that the lower or inner surface of the vitelline membrane is specially constructed to support spreading. This may be so, but it is not the only surface that will do so. A cellulose-ester polypore filter will serve as well.

The tautness of the membrane is also important. A flaccid vitelline membrane does not support spreading, whereas a local increase in tautness is associated with a local increase in spreading. This raises the possibility that stress lines in the membrane may guide the outward spreading of the blastoderm. This hypothesis does not stand the test, however. If a blastoderm is inverted so that the hypoblast is in contact with the vitelline membrane, it will curl under so that its ectoderm is once again in contact with the membrane. Then it will spread backward toward the animal pole (Fig. 13-1).

Like other spreading cell sheets the epibolic spreading of the blastoderm depends primarily on strong adhesion at its margin. The rest of the blastoderm is either nonadherent or lightly adherent. Unlike some other spreading cell sheets, however, the only cells on the chick blastoderm that will spread are those *normally* at the margin. Apparently the marginal cells are intrinsically different. When examined in the electron microscope they

Fig. 13-1 Behavior of chick blastoderm explanted normally onto vitelline membrane (a). Adhesive surface attaches to the vitelline membrane (b). Normal expansion follows as shown by diagram of whole preparation (c). (d), (e), (f), behavior of blastoderm inverted on vitelline membrane. Blastoderm edge curls under to bring adhesive surface against vitelline membrane (e). Expansion in this case results in formation of a hollow vesicle (f). (In c and f, thick line denotes ectoderm, broken line endoderm. In a, b, d, and e, shading indicates adhesive surface.) (New. 1959. J. Embryol. Exptl. Morphol. 7:146.)

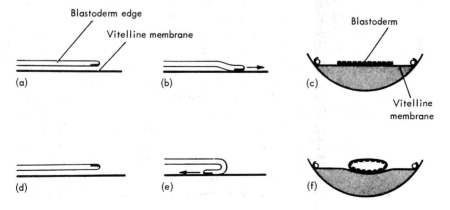

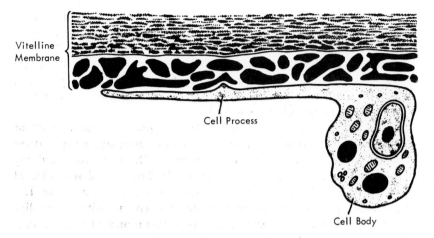

Fig. 13-2 Diagram of a cell at the margin of overgrowth of a chick blastoderm. The cell process is represented here as shorter than in the specimens. (Bellairs. 1963. J. Embryol. Exptl. Morphol. 11:201.)

are found to extend long thin processes out onto the vitelline membrane. These processes may be enormously long (as much as 500 μ!). They may be as little as $\frac{1}{4}\mu$ deep (Fig. 13-2). Only marginal cells extend these processes. Apparently they are the locomotor organs. It is suggested that they probably contract and pull the marginal cells with them. Significantly, the margin of the blastoderm adheres so tightly to the vitelline membrane that cells and membrane remain attached throughout the fixation procedure. Projections from the upper surface of a process are observed to extend up into the vitelline membrane and seem to anchor the process to the membrane. Like other hyaline extensions from cells, these processes lack mitochondria, Golgi apparatus, endoplasmic reticulum, and yolk droplets. It is evident that these marginal cells are most unusual. They deserve further intensive study, in cell culture as well as in situ.

The possession of these long processes suggests that the marginal cells are the prime movers in blastoderm epiboly. This hypothesis is supported by two additional lines of evidence: 1) blastoderms expand only when their margins are spreading, and 2) the marginal region of a blastoderm will spread even though isolated from the rest of the blastoderm. In this regard the spreading chick blastoderm appears to resemble a spreading epithelial sheet, and the same questions may be raised concerning the activities of the nonmarginal cells. We have no direct information on the degree of passivity or activity of the nonmarginal cells, but it is established that they are put under much tension by the spreading margin. This suggests that to some degree at least they are passively pulled by the margin. Since the blastoderm is placed under tension during epiboly the nonmarginal cells must be tightly adherent, or they would be pulled apart. It is significant, therefore, that the ectodermal cells, which form a cohesive sheet throughout epiboly, are joined to each other only by focal tight and close junctions after an 18-hour period of incubation, when the ectoderm is not yet greatly stretched. By three days, however, when the stretching is more intense, these junctions are replaced by zonal tight junctions (*zonulae occludentes*), *zonulae adhaerentes,* and desmosomes (*maculae adhaerentes*). In the latter two the opposed plasma membranes are 250 A apart.

It was once popular to suppose that the spreading of a sheet is due to centrally located centers of proliferation that cause spreading by "pushing from behind." Evidence from the chick blastoderm gives no support to such a proposal. Cell division goes on during epiboly, to be sure, but it is primarily near the margin of the blastoderm. Moreover, it is not necessary for blastoderm expansion. A blastoderm will continue to spread after mitoses have been completely inhibited by a folic acid antagonist. Finally, as Schlesinger has shown, marginal regions of the blastoderm will spread even though isolated from the rest of the blastoderm.

Endoderm and mesoderm formation

The blastoderm of an unincubated hen's egg is composed of a cohesive epiblast underlain by a ring of loosely packed hypoblast cells. The hypoblast is the precursor of part of the endoderm. Spratt and Haas, using carbon marking, have shown that during the first few hours of incubation the hypoblast moves forward and radially on the undersurface of the epiblast from a thickened posterior region. The undersurface of the epiblast appears to be a specific substratum, since hypoblast cells have not been found to migrate on the upper surface of the epiblast, the inner or outer surface of the vitelline membrane, or on agar or glass. But the epiblast does not orient or give direction to hypoblast movement. By merely changing the position of its thickened posterior region, the hypoblast can be forced to traverse the epiblast in any direction.

What then gives the hypoblast cells direction? And what causes them to begin their migration? Spratt and Haas believe that intensive proliferation in the thickened posterior part of the hypoblast is the cause of both. Cells are thought to commence movement passively as a result of a push due to increased population density of the rapidly dividing growth center and to continue movement anteriorward because of a continuation of this vis a tergo. This hypothesis rests on two observations. Hypoblast cells stop moving in the anterior part on the blastoderm if they are isolated from the posterior half, and they begin movement in any part of the blastoderm if a proliferation center forms among them. This is an interesting idea, though not a new one. Cellular proliferation has often been suggested as the motive force for mass cell movements. Generally, however, when subjected to analysis, such notions have not fared well. Nevertheless, each case deserves examination in its own right.

The proposal that hypoblast cells are pushed passively by a burgeoning growth center has not yet been adequately tested. No mitotic counts or other evidence of cell division have been submitted to demonstrate that the posterior part of the hypoblast is in fact a proliferation center. One worker, Vakaet, finds no evidence for proliferation and proposes that the hypoblast forms primarily by polyinvagination, with cells slipping from the upper to the lower layer at many points. If there is much proliferation in the posterior region, obstruction of the anteriorward migration should cause a piling up of cells. Blockage of various sorts causes no such accumulation. The most crucial test of whether proliferation is essential for a process is to apply a mitotic inhibitor and see if the process is blocked. When this was done for the spreading chick area opaca and *Fundulus* blastoderm (see p. 189), epibolic spreading continued regardless. Mitotic inhibitors have not been applied to chick blastoderms during the period of hypoblast migration. In consequence, we do not yet know whether the proliferation center hypothesis is valid or not.

While awaiting an adequate test of this hypothesis it seems more fruitful to adopt a working hypothesis that gives a more active role to the migrating hypoblast cells. In systems that are well analyzed, the cells of the leading edge invariably play an active role. In the chick blastoderm, the accumulation of hypoblast cells in the posterior region may merely provide a source of cells for the ensuing cell movements, much as an explant is the source of cells of the zone of outgrowth in a tissue culture. It is possible that movements start as a result of changes in cell surface activity, such as motility and adhesiveness. They could, for example, become specifically contact inhibiting, such that, like fibroblasts in tissue culture, they are inhibited from moving over each other but will move over the alternative substratum, in this case the undersurface of the epiblast. As for the directionality of their movement, hypoblast cells are bound to move anteriorward; they have nowhere else to go.

No explanation has been provided for the cessation of movement of hypoblast cells in an isolated anterior half of a blastoderm. If type-specific contact inhibition is at work, it is possible that these cells require frequent contact with other hypoblast cells in order to acquire the impulse to move. Abercrombie and Gitlin have shown that if a given cell moves off so that it becomes isolated from the outgrowth, its movement soon ceases, probably because contact inhibition of movement serves to stimulate movement away from the contact. But as soon as the contact is lost the stimulus is gone. Isolation of hypoblast cells from the main mass of the hypoblast in the posterior half of the blastoderm would deprive them of this contact and perhaps therefore of their main stimulus to move.

It has been established for some time now by detailed marking studies with vital dyes and carbon that the primitive streak forms as a result of convergence of epiblast cells to the midline of the blastoderm. There they sink in and migrate out laterally between the epiblast and the hypoblast to form the mesoblast. Rosenquist has reinvestigated this problem recently by labeling epiblast cells with tritiated thymidine and grafting them to unlabeled blastoderms. These experiments have confirmed the previous studies and shown that some endoderm also originates from the epiblast by invagination at the primitive streak. At this time, the epiblast is a simple pseudostratified columnar epithelium, whereas the hypoblast is a simple squamous epithelium. The mesoderm is a mesenchyme, in the sense that the cells are more loosely organized and have no free surface. Mesoblast cells derive from the basal surface of the epiblast in the region of the primitive streak (Fig. 13-3).

Fig. 13-3 Diagram of a four somite (stage 8) chick embryo sectioned through the primitive streak. In the inset the streak and the forming mesoblast cells are shown at higher magnification. (Courtesy of E. Hay. Hay. 1968. Epithelial-mesenchymal interactions. Williams & Wilkins, Baltimore. p. 31.)

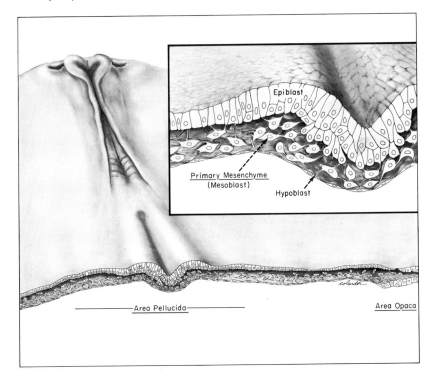

Studies on the mechanism of formation of the mesoblast thus far have been confined to observation of cell relationships at the level of fine structure. Although these necessarily deal entirely with fixed material, they have nevertheless yielded important leads, based as they are on the previous marking studies. Early investigations (back around 1960) showed that desmosomes and so-called "terminal bars" (*zonulae adhaerentes* and *zonulae occludentes*) do not appear in any of the layers during the primitive streak stage (see p. 60). In addition, these studies revealed the presence of bottle or flask-shaped cells in the primitive streak at all levels. Balinsky and Walther were fascinated by the obvious possibility that this is the manner in which epiblast cells invaginate and postulated that this change in cell shape is brought about by contraction of an electron-dense layer found in the cytoplasm of the neck region of flask-shaped cells. Granholm has observed microtubules in the necks of flask cells and has postulated that they may aid elongation of these cells. Both of these observations are interesting but the postulations await experimental proof. Detailed study of cell shape changes and the junctional relations of the opposed plasma membranes awaited the resolution provided by contemporary fixation and staining methods. Such a study has just been made by Trelstad, Hay, and Revel. The ensuing discussion will be based on their observations.

Lateral to the primitive streak and in the streak itself the epiblast cells are in contact by means of numerous minute tight junctions or points of apparent fusion. These are punctate or focal junctions and not at all like the belts of *zonulae occludentes* seen in more advanced epithelia. There are also focal close junctions with a space of 20–100 Å between opposed cell membranes. The fine structure of these cells gives no indication as to how they converge toward the streak, and this remains a crucial unsolved problem. Isolation experiments suggest that the place to look is within the epiblast itself. The orientation of the convergence is apparently controlled by the blastoderm that surrounds the streak and not by the primitive streak.

The stellate mesoblast cells that leave the basal layer of the epiblast at the primitive streak are distinctly polarized and at first flask shaped. The basal part of the cell contains the nucleus and a paucity of organelles, and the apical part the Golgi complex (like cells of the epiblast). As each cell leaves the epiblast its basal part leads off, and the apical end, containing the Golgi complex, trails behind. The leading end and, to a lesser extent, the lateral regions develop filopodia. These filopodia contain filaments about 50 Å in diameter, which could conceivably have a contractile function. They do not contain microtubules. But microtubules are common in the cell bodies, where they may play a role in shaping cells and in cell movements. Broader projections are also evident at the leading edges of mesoblast cells. These may correspond to ruffled membranes. All mesoblast cells are connected to each other and to cells of the primitive streak by focal tight and close junctions, especially along their trailing surfaces (Fig. 13-4). These observations dispel the notion that mesoblast cells break away from the primitive streak and migrate as individuals. They form a loosely connected network—no doubt in part due to their junctional contacts. Some of these junctions may be formed anew, but it is also possible that some represent original sites of contact in the streak that are retained in the mesoblast. More extensive tight junctions, involving greater areas of the cell surface, do not appear in the mesoblast until after gastrulation, when it is engaged in forming tissues.

Filopodia extending from the advancing edge of the uppermost and lowermost mesoblast cells also attach to the adjacent epithelial surfaces of the epiblast and the hypoblast by means of focal tight and close junctions. Connections are made both with the incomplete basal lamina of the epithelia and with bare areas of the plasma membrane, but they seem to favor the latter. It seems likely that these contacts are used for traction while the mesoblast cells crawl out between the two epithelial layers. They are therefore no doubt of a temporary nature. Cells midway between the epiblast and hypoblast have the most numerous filopodia, often thrusting them out freely into intercellular space. These filopodia contact other mesoblast cells and may also be involved in cell movement. Hay postulates that the tight junctions made by the mesoblast filopodia are contact inhibiting. This seems unlikely, at least relative to the epiblast and the hypoblast, since these cells appear to be able to move readily along their surfaces. The essence of contact inhibition is lack of movement over other cells. Of course, if the mesoblast cells were contact inhibiting in a specific way, i.e., toward cells of the same type, then they would be inhibited from crawling over each other and would choose instead alternative substrata, which in this instance would be the epiblast and the hypoblast. In any case, no decision can be made until direct observations have been made of the locomotion of living mesoblast cells, both within the blastoderm and in culture. This will be difficult to do, but it is essential if we are to understand the significance of the cell contacts and shape changes revealed by the electron microscope.

These fine-structural studies have been accompanied by electrophysiological studies of coupling between cells at later stages of development. Electrical current fed into a cell in the medullary plate, for example, can be picked up in a cell in the presumptive notochord. Tight junctions between cells have been implicated by a number of workers (see Ch. 4, p. 59) as probable sites of low electrical resistance (although it is also possible in some instances that current may pass by way of the intercellular fluid). Where such electrical coupling is demonstrated it is always possible that there is preferential ion flow between the cells involved. This could

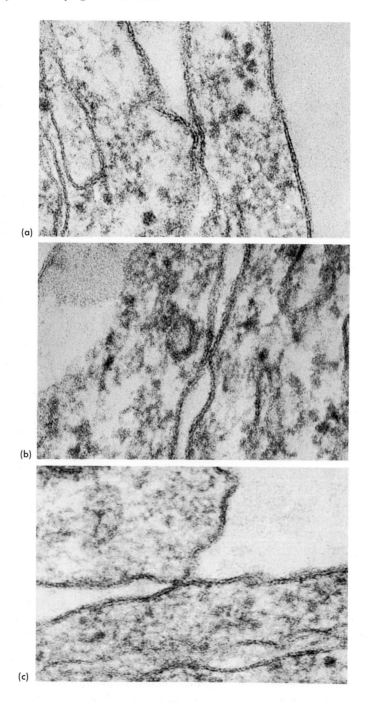

have important implications for cell movements, if such ion exchange has an effect of the surface activity of cells, such as on cytoplasmic contractility. Hay et al. speculate (as have others) that in such a manner electronic coupling could control contact inhibition. By this means, they suggest that coupling could control the movement of the mesentoblast sheet during chick gastrulation. Although such a proposal has much heuristic value and may turn out to have some validity, it must be emphasized that at present it is almost without factual basis. The proposed relation between contact inhibition and electrical coupling, although much discussed, has not been established yet for any system; moreover, the requisite electrical measurements have not yet been made of the relations between the cells under discussion (mesoblast, epiblast, and hypoblast of the primitive streak stage); and, finally, nothing is known about the possible contact inhibitory behavior of chick mesoblast cells, except by extrapolation from electron micrographs. (For further discussion of this matter, see p. 208.)

Migration of precardiac mesoderm

Between the primitive streak stage and the first appearance of the somites, precardiac mesodermal cells migrate from lateral paired precardiac regions to the midline site of heart formation. DeHaan studied this process with time-lapse cinemicrography and found it to be exceptional in that it involves neither the migrations of separate cells nor the spreading of cell sheets but the movement of clusters of cells. (Actually a recent study of this process with precardiac cells labeled with tritiated thymidine has shown that the cells observed with time-lapse are really *preendocardial* cells. The *preepimyocardial* cells are formed in a sheet of splanchnic mesoderm and move medially as a sheet, like most cells engaged in morphogenetic movements.)

The preendocardial mesoderm first aggregates into small cell clusters which adhere to the endoderm. These clusters then migrate independently of one another over the endoderm eventually to reach the midline. At first, their movements are random, bearing no relation to their eventual anteromedial point of convergence. After formation of the head fold, the clusters begin to follow oriented and parallel pathways toward their destination (Fig. 13-5).

Fig. 13-4 High-magnification electron micrographs showing junctions between dissimilar cells in a definitive primitive streak blastoderm (stage 4) near the streak. (a) Close junction between plasmalemmas of mesoblast and presumptive endodermal cells (hypoblast). (b) Tight junction between mesoblast and epiblast. (c) Tight junction between mesoblast and hypoblast. × 280,000. (Courtesy of Elizabeth Hay. After Trelstad, Hay, and Revel. 1967. Develop. Biol. **16**:78.)

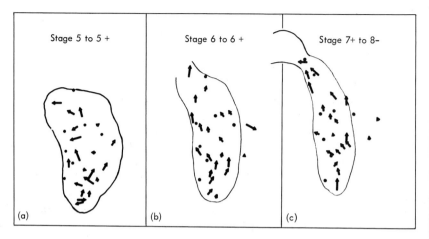

Fig. 13-5 Active migration of clusters of precardiac mesoderm cells of a chick embryo during selected 1-hour periods. (a) Movement during the head-process stage (5–5+). (b) Movement during the head-fold stage (6–6+). (c) Movement during formation of the first somites (7+ –8–). (DeHaan. 1963. Exptl. Cell Res. **29**:544.)

Since the clusters move over the endoderm, it seemed possible that this layer might give orientation to the clusters. Examination of the endoderm revealed that at the head process stage the endoderm cells lose their irregular polygonal shape and assume a spindle form, oriented in the direction of the site of heart formation. The coincidence of time and place is striking and suggests that the oriented lines of endodermal cell junctions (or other linear structures at the cell surface that become aligned with the alteration of cell shape) may give orientation to the mesodermal structures. Since this would suffice only to give orientation and not direction, DeHaan sought a supplementary mechanism. He speculated that if the endoderm were increasingly adhesive in an anteriorward gradient, the cells might adhere more readily on their anterior sides and would then tend to move up the gradient toward the point of heart formation (in a fashion subsequently demonstrated for L-cells in culture by Carter). This proposal, though not improbable, is of course completely hypothetical. Appropriate measurements of endodermal adhesiveness have not yet been made.

Whatever may be the mechanism of giving direction to cluster translocation, it is essential to know just how the clusters move. Presumably their movement is due to the activities of their constituent cells. In this connection it is pertinent to contrast the mobility of preendocardial clusters with the immobility of clusters of retinal pigment cells (p. 119). The immobility of pigment cell clusters correlates with the very low mobility of individual pigment cells. Perhaps the high mobility of precardiac clusters correlates with a correspondingly high mobility of individual precardial cells. This would be an interesting point to establish, for, if confirmed, it would constitute evidence that the movement of a cell cluster does indeed depend on the collective migratory activities of its component cells.

SELECTED REFERENCES

Balinsky, B. I. and H. H. Walther. 1961. The emigration of presumptive mesoblast from the primitive streak in the chick as studied with the electron microscope. Acta Embryol. Morphol. Exptl. **4:**261-283.

Bellairs, Ruth. 1963. Differentiation of the yolk sac of the chick studied by electron microscopy. J. Embryol. Exptl. Morphol. **11:**201-225.

DeHaan, R. L. 1963. Migration patterns of the precardiac mesoderm in the early chick embryo. Exp. Cell Res. **29:**544-560.

Hay, Elizabeth D. 1968. Organization and fine structure of epithelium and mesenchyme in the developing chick embryo, p. 31-35. *In* R. Fleischmajer and R. E. Billingham [ed.] Epithelial-mesenchymal interactions. Williams & Wilkins, Baltimore.

New, D. A. T. 1959. Adhesive properties and expansion of the chick blastoderm. J. Exptl. Embryol. Morphol. **7:**146-164.

Overton, Jane. 1962. Desmosome development in normal and reassociating cells in the early chick blastoderm. Develop. Biol. **4:**532-548.

Rosenquist, G. C. 1966. A Radioautographic study of labeled grafts in the chick blastoderm. Development from primitive streak stages to stage 12. Carnegie Inst. Wash. Publ. 625:Contrib. Embryol. **38:**71-110.

Rudnick, Dorothea. 1955. Teleosts and birds, p. 297-314. *In* B. H. Willier, P. Weiss and V. Hamburger [ed.] The analysis of development. Saunders, Philadelphia.

Schlesinger, A. G. 1958. The structural significance of the avian yolk in embryogenesis. J. Exptl. Zool. **138:**223-258.

Spratt, N. T. Jr. 1946. Formation of the primitive streak in the explanted chick blastoderm marked with carbon particles. J. Exptl. Zool. **103:**259-304.

Spratt. N. T. Jr. 1963. Role of the substratum, supracellular continuity, and differential growth in morphogenetic cell movements. Develop. Biol. **7:**51-63.

Trelstad, R. L., E. D. Hay, and J. P. Revel. 1967. Cell contact during early morphogenesis in the chick embryo. Develop. Biol. **16**:78–106.

Vakaet, L. 1960. Quelques précisions sur la cinématique de la ligne primitive chez le poulet. J. Embryol. Exptl. Morphol. **8**:321–326.

Waddington, C. H. 1952. The epigenetics of birds. Cambridge Univ. Press, Cambridge. 272 p.

FOURTEEN

Teleost Epiboly*

Teleost eggs have attracted attention for the analysis of gastrulation because of their transparency and the spectacular extent of their epiboly. The egg is highly meroblastic, with cleavage occurring only in the blastodisc that surmounts a fluid yolk sphere of varying size. When the blastoderm has divided into a few thousand cells, it flattens and slowly spreads over the yolk finally to encompass it completely with closure of the "blastopore" (Fig. 14-1). While this is going on, the margin of the blastoderm thickens to form the so-called germ ring. The cells of the germ ring then converge toward the middorsal line to form the embryonic shield. Apparently there is no involution at the margin of the blastoderm. The embryonic shield increases in mass and elongates as epiboly progresses and eventually gives rise to the embryo. The rest of the blastoderm forms the yolk sac, which later becomes vascularized. During the blastula and early gastrula stages the blastoderm consists of two general classes of cells (Fig. 14-2). Those at the surface of the blastoderm are organized into a cohesive layer one cell thick called the enveloping layer or Deckschicht. This layer engages fully in epiboly but does not give rise to any tissues of the embryo. Beneath and inside the enveloping layer are the so-called deep cells. These cells engage in extensive morphogenetic movements, including formation of the germ ring and the embryonic shield, and eventually give rise to the embryo. The normal course of the movements of these cells has recently been subjected to detailed analysis by Ballard. Beneath the

*I hope I may be forgiven if rather more attention is devoted to teleost epiboly than to morphogenetic movements in other groups. It is a subject which has occupied me off and on for some years, and I know it with the kind of intimacy that only comes with direct personal experience. I feel therefore that I have more to say.

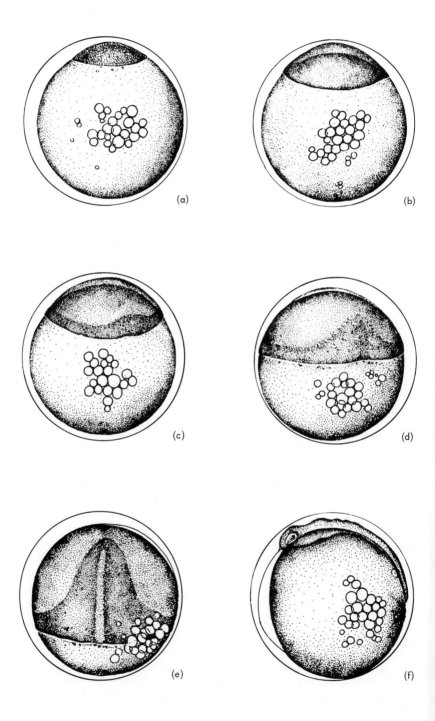

Fig. 14-1 Normal development of the teleost of *Fundulus heteroclitus*. (a) Midblastula. The many-celled blastoderm surmounts the large yolk sphere. The spherical bodies in the yolk are lipid droplets. The periblast is not shown. (b) Beginning gastrula. The blastoderm has flattened on the yolk and is beginning to spread. (c) Early gastrula. The blastoderm has already spread somewhat over the yolk and its marginal region has thickened to form the germ ring. Cells within the germ ring are converging dorsad to form the embryonic shield. (d) and (e) illustrate mid to advanced gastrulae. Elongation of the shield accompanies epiboly. (f) Closure of the blastopore. The embryo is forming out of the embryonic shield anterior to the point of blastopore closure. Note the optic vesicles. (Armstrong and Child. 1965. Biol. Bull. **128**:143.)

---

enveloping layer and deep cells and separating them from the fluid yolk is a curious and interesting layer, the periblast. It is a syncytium which arises during early cleavage stages by incomplete division of the peripheral and basal cytoplasm. It is invaded by nuclei from peripheral blastomeres. These nuclei continue to divide for a short time, populating the cytoplasmic layer and converting it into a syncytium. The periblast undergoes epiboly along with the blastoderm and, as we shall see, plays an important part in the process.

Research on the mechanism of epiboly in teleost eggs has been pursued mainly in three laboratories, that of Warren L. Lewis in Philadelphia, Charles Devillers in Paris, and in my own laboratory in New Haven and Woods Hole. It was Lewis who really began the work and elaborated the first hypothesis. (It was one of his last research projects in a long and

Fig. 14-2 Diagram of an early gastrula stage of *Fundulus*. The blastoderm is flattened and extends over one-third of the yolk as a result of epiboly. The cells of the enveloping layer (EL) are flattened and closely united. Lobopodia extend from some of the deep blastomeres (DB). Note that the periblast (P) and the yolk cytoplasmic layer (YCL) are parts of a continuous cytoplasmic layer. The periblast is the thicker nucleated (PN) part which lies between the blastoderm and the yolk (Y). SC, segmentation cavity. The clear area immediately beneath the periblast and yolk cytoplasmic layer is a region where yolk is apparently being digested. ×140. (Lentz and Trinkaus. 1967. J. Cell Biol. **32**:121.)

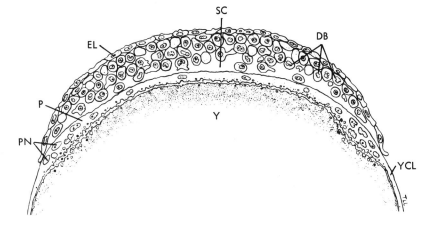

distinguished career.) He worked mainly on the egg of the zebra fish, *Brachydanio rerio,* and was soon impressed by the remarkable properties of the surface layer of the yolk. His observations were confirmed and extended by Devillers for the trout egg and by myself for the egg of *Fundulus.*

The yolk surface layer is easily perceived because of its exceptional thickness and sturdiness. Electron micrographs reveal it to be approximately 0.6 μ thick. It connects to the margin of the blastodisc and early blastoderm and later to the margin of the periblast as the latter extends peripheral to the blastoderm. Its outer surface is nonadhesive and its inner surface sticky. When the yolk surface layer is punctured, the wound widens within seconds and becomes round, demonstrating that the membrane is under considerable uniform contractile tension. Then, as the edges of the wound thicken to form a gelated ring, the wound closes in a matter of minutes, apparently as a result of the contraction of the gelated ring. As it closes, the surrounding surface layer is stretched. When a wound is made near the margin of the blastoderm, the stretching extends to it; the blastoderm margin is pulled toward the point of wound closure; and its blastomeres are stretched into a spindle form (Fig. 14-3). During wound closure, the shape of the egg becomes distorted, but soon after closure the semispherical shape of the egg is restored, presenting additional evidence that the yolk surface layer exerts contractile tension tangentially and uniformly in all directions. Still further evidence is the fact that two

Fig. 14-3 Stretched cells at the edge of a *Fundulus* blastoderm resulting from wound closure in the surface layer of the yolk sphere in an early blastula stage. SC, stretched cells; WC, wound closure; F, folds in surface layer of yolk. ×100. (After Trinkaus. 1949. Proc. Natl. Acad. Sci. **35:**218.)

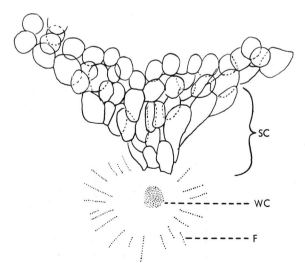

or three wounds in different parts of the yolk surface layer of the same egg will widen simultaneously.

On the basis of such evidence Lewis proposed that the cortical layer of the yolk is the prime mover in epiboly. The hypothesis assumes that before epiboly begins the contractile tension of the yolk layer is balanced by that of the blastoderm and periblast but that a decrease in the contractile tension of these layers upsets the balance. In the latter case, contraction of the yolk layer then pulls the blastoderm over the yolk to the vegetal pole. The yolk surface layer is thought also to exert pressure on the yolk mass, compressing it against the periblast and blastoderm and expanding them. Lewis thinks of the yolk mass acting as a sort of hydrostatic cushion. As the yolk layer contracts, he assumes that its inner aspect solates into yolk endoplasm.

Observations on all three eggs make it clear that the yolk surface layer does indeed exert contractile tension on the margins of the blastoderm and syncytial periblast. It also compresses the fluid yolk into a spherical shape. But this is not sufficient evidence for Lewis' hypothesis. It requires a demonstration that the yolk layer actually exerts sufficient contractile tension to pull the blastoderm over the yolk in epiboly and that this contractile tension is indispensible for epiboly. There is no information on the amount of tension exerted by the yolk cortex, other than what has already been described, and this is not enough to tell whether or not the tension is adequate. As far as the second issue is concerned, there is much evidence that epiboly may occur in the absence of a connection to the yolk cortex. In *Fundulus,* the periblast begins to spread prior to the blastoderm, and as a consequence the marginal periblast widens greatly during late blastula stages. Then the blastoderm spreads over the periblast until it reaches the periblast margin, toward the end of the first phase of epiboly (Fig. 14-4). Similarly, at the end of gastrulation the periblast moves ahead of the blastoderm and closes its blastopore prior to the latter. Clearly blastoderm epiboly begins and ends normally without a connection to the yolk surface layer. This phenomenon can be repeated experimentally by severing the marginal connection of the blastoderm to the periblast. The blastoderm retracts immediately, then slowly reattaches to its periblast substratum and resumes its expansion (Fig. 14-5). When performed in early to middle stages of epiboly, the blastoderm spreads fast enough to catch up to the margin of the spreading periblast and then moves ahead normally at the same rate as the periblast. When performed after about three-fourths of epiboly has occurred, the blastoderm retracts, as always, but the periblast behaves differently. Its margin contracts into the yolk with such force that it closes its blastopore in a few minutes and pinches off the remaining uncovered yolk. As the yolk is constricted contractile tension in the yolk cortex tends to oppose constriction of the periblast.

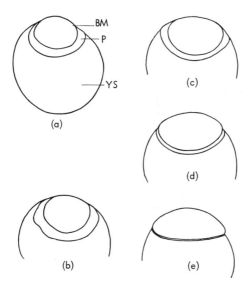

Fig. 14-4 Epiboly of a *Fundulus* blastoderm and periblast, late blastula to beginning gastrula. Note the decrease in width of the marginal periblast during the first stages of blastoderm expansion [from (b) to (e)]. Time intervals since (a) are: (b) 175 minutes; (c) 245 minutes; (d) 365 minutes; (e) 450 minutes. BM, blastoderm; P, periblast; YS, yolk sphere. (After Trinkaus. 1951. J. Exptl. Zool. 118:269.)

The latter continues anyway, demonstrating that its force is greater than that of the yolk cortex. With closure of the periblast blastopore, the yolk surface layer ceases to exist. Nevertheless, the blastoderm readheres to the periblast and moves over it until it closes its own blastopore.

Since the blastoderm may undergo epiboly even though it lacks a connection to the yolk surface layer, the contractile tension of the yolk cortex is obviously not necessary. In consequence, the yolk-surface-layer hypothesis is not supported by the facts and has lost its usefulness as a guide to the study of teleost epiboly. It should be noted that Devillers does not share this opinion. On the basis of essentially the same evidence he considers the yolk surface layer to be "a supplementary force which is not negligible" (1961). This is of course possible, but he offers no new evidence.

The experiments that led to the demise of the yolk-surface-layer hypothesis directed attention to certain remarkable properties of the blastoderm and the periblast. The blastoderm appears to have an intrinsic capacity to spread, using the expanding periblast as its substratum. The periblast is an ideal substratum for the blastoderm because it also spreads in epiboly. Moreover, it always underlays the blastoderm during epiboly and even at times extends beyond it. In addition, the periblast can engage in epiboly independently of the blastoderm. If a periblast is completely denuded of its blastoderm at an early gastrula stage, it will spread the entire distance over the yolk finally to engulf it completely. Evidence that an expanding periblast substratum is essential for blastoderm epiboly is provided in a pretty experiment of Devillers. If an early gastrula blastoderm of the trout egg is grafted to an exposed periblast of the same stage, the blastoderm adheres and spreads over the expanding periblast. But if an early gastrula blastoderm is grafted to a periblast of the blastula stage, before the periblast has begun to spread, no epiboly occurs. The blastoderm expresses its intrinsic capacity to spread, however, and pushes up a perpendicular protuberance (Fig.14-6). A similar dependence on the expanding periblast substratum is found in *Fundulus*. A reattached blastoderm expands faster than the periblast up to the margin of the latter, but then it slows down to the rate of the periblast. It never pushes beyond to move over the yolk surface.

The capacity of the blastoderm to spread due to forces within it is not dependent on the blastoderm being intact. Removal of either the dorsal or the ventral half of the blastoderm in *Fundulus* results in an increased expansion of the other half. Indeed, as long as there is some uncovered periblast the remaining part of the blastoderm will strive to cover it. Sometimes its expansion is great enough to cover the yolk entirely, a dramatic demonstration that the capacity of parts of the blastoderm for spreading is much greater than is normally expressed. As far as could be determined, ventral and dorsal halves have about the same capacity to spread. This is consistent with the results of carbon marking, which showed that the blastoderm spreads rather uniformly during normal epiboly (except for greater marginal expansion dorsally). In the trout (and perhaps in large teleost eggs in general) the ventral region spreads much more than the dorsal. In normal epiboly this displaces the blastopore in a dorsal direction.

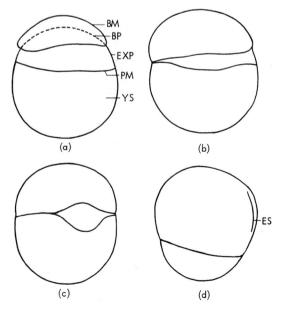

Fig. 14-5 Reexpansion of a reattached blastoderm of an early gastrula over the periblast in *Fundulus*. Time intervals since severance of the marginal connection of the blastoderm to the margin of the periblast: (a) 30 minutes; (b) 105 minutes; (c) 330 minutes; (d) 480 minutes. BM, blastoderm; BP, periblast underneath blastoderm; ES, embryonic shield; EXP, exposed periblast; PM, margin of periblast; YS, yolk sphere. (Modified from Trinkaus. 1951. J. Exptl. Zool. **118**:269.)

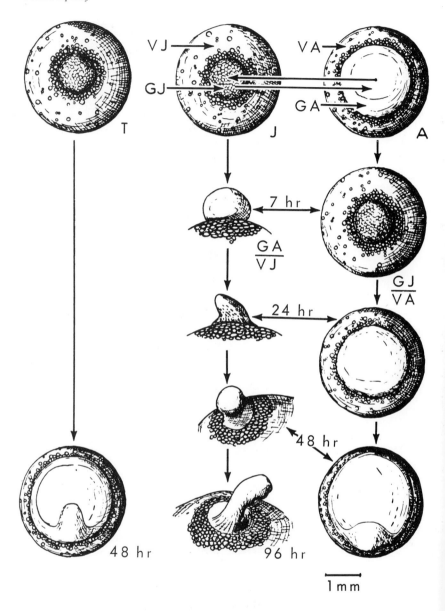

Fig. 14-6 Experiments on the coordination of the epibolic forces in the egg of *Salmo irideus*. Left column: evolution of the (young) control; central column: combination of an old germ GA (coming from a gastrula of the right column) and a young yolk VJ; right column: combination of a young germ GJ (coming from the central column) and an old yolk. In this combination epiboly only starts again after 24 hours at the same time as in controls. Note the difference of development between the normal embryo (left) and that of the composite germ at the same moment (96 hr). (Devillers. 1960. Advan. Morphogenesis. **1**:379.)

The spreading of parts of the blastoderm focuses attention on the cellular constituents of the blastoderm. In the trout egg spreading activity is apparently confined to the outer cohesive cell layer one cell thick called the enveloping layer (Fig. 14-2). Pieces of this layer from gastrulae spread in culture, whereas explants of the inner cells do not. In *Fundulus* it has not yet been possible to separate the enveloping layer from the rest of the blastoderm. Our first approach was to dissociate blastoderms into cell suspensions with EDTA and culture them in standing drops on glass in a simple nutrient medium. Under these conditions blastula cells remain spheroid and protrude lobopodia. Early gastrula cells, on the other hand, flatten and spread extensively on the glass within an hour or two after culturing (Fig. 14-7). This correlation of cell spreading in vitro with the onset of epiboly indicates that epiboly of the blastoderm is due in part to the spreading activities of its constituent cells. Late blastula cells flatten more extensively with increased time in culture, when control eggs are beginning epiboly. This suggests that the process which changes the character of the cells within the blastoderm continues in culture, independently of the change in environmental conditions. The flattening of

Fig. 14-7   1) *Fundulus* blastula cells; 29 minutes in culture. Note lobopodia. No cells flattened. 2) Same cells as in 1; 5 hours in culture. Note a small amount of aggregation. A few cell clusters slightly flattened. 3) Early gastrula cells; 18 minutes in culture. Cells already flattened. 4) Same cells as in 3; one hour in culture. Note much aggregation. Cells extensively flattened. (Trinkaus. 1963. Develop. Biol. **7**:513.)

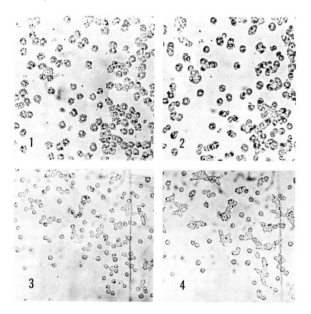

Fundulus gastrula cells recalls the flattening of amphibian ectoderm and endoderm cells in culture. Of course, this increased flattening of gastrula cells could be due to several factors (see Ch. 2). As a working hypothesis, we have proposed that it is due at least in part to increased adhesiveness of the cell surface. If only one surface is available, as in cell cultures on a glass substratum, adhesive cells will tend to flatten on it, with larger adhering areas producing greater degrees of flattening. Conversely, the more rounded form of blastula cells in culture may be a reflection of lesser adhesiveness. As already emphasized (p. 26), the motility of cells is necessarily related to their adhesiveness. If surface adhesiveness is too low, as appears to be the case for *Fundulus* blastula cells, a cell will not be able to get a grip on the substratum and pull itself over it. Thus an increase in adhesiveness of gastrula cells, over and above the low adhesiveness of blastula cells, could well be necessary for the movements of gastrula cells in epiboly.

But this is certainly not the whole story. It has already been pointed out that cell motility depends on at least two properties: a high enough degree of adhesiveness (but not too high) and possession of an active organ of locomotion, such as a ruffled membrane. A highly active ruffled membrane could also aid flattening by causing a cell to crawl out on and stretch itself over the substratum. Thus both properties could jointly contribute to the flattening and spreading of cells. In our studies of *Fundulus* gastrula cells in culture we have not yet distinguished between these properties.

We may conclude at this point that the blastoderm and its constituent cells acquire the intrinsic capacity for spreading during late blastula stages and that this spreading capacity is expressed during gastrulation by epiboly over its natural substratum, the expanding periblast. From what is known about the spreading of other cell sheets it comes as no surprise to learn that the manner of adhesion of the blastoderm to its substratum is critical. When the marginal connection between the enveloping layer and the periblast is severed during epiboly (Fig. 14-5), the blastoderm retracts promptly, showing both that it is under tension and that marginal cells adhere more strongly to the periblast than do nonmarginal cells. This recalls other spreading epithelial sheets and suggests that the marginal cells may be the prime movers in epiboly, exerting tension on the blastoderm by their outward spreading. In this regard, the behavior of *Fundulus* blastoderms which have been isolated from the yolk and periblast is significant. They roll up and lack a free edge. Under such circumstances the enveloping layer spreads little if at all. The spreading of the marginal cells of the enveloping layer is clearly important for blastoderm expansion. It should be emphasized, however, that a capacity for adhering strongly to the periblast and spreading over it is not an exclusive property of cells which are normally at the margin (as is apparently the case for the chick blastoderm). When part of a blastoderm is removed, the cells of the free cut surface adhere to the periblast. Thus, when central cells become marginal they behave like marginal cells. The partial blastoderm now spreads over the exposed periblast. (Although the marginal cells of the enveloping layer are the only cells which normally adhere *strongly* to the periblast, they are not the only ones to adhere. The lowermost deep cells also adhere somewhat, as shown by a tendency of some of them to remain stuck to the periblast when a blastoderm is removed. The degree of adhesiveness appears, however, to be low. They are readily brushed off with a hair loop.)

It has often been postulated (see Chs. 12 and 13) that embryonic sheets could be caused to spread or fold as a result of localized increases in the rate of cell division. The best way to test this hypothesis is to block mitoses and see if the process continues. Kessell did this in *Fundulus* with colchicine and found that epiboly continues anyway, in the complete absence of cell division. It may be concluded that even though cell divisions go on during epiboly they are not essential for it. In view of the fact that colchicine blocks mitosis by inhibiting assembly or by causing disassembly of the microtubules of the mitotic spindle (see p. 207), the results of this experiment render it unlikely that microtubules play an essential role in *Fundulus* epiboly. Significantly, microtubules have not been seen in normal blastomeres, except in the mitotic spindle.

At this juncture it became clear that further knowledge of the surface activities of the blastoderm cells in situ and their contact relations with each other and with the periblast was necessary. For this reason we began a correlated time-lapse cinematographic and fine-structural study of cell surface activities and cell contacts within normally developing blastoderms. The picture which emerges from the cinematography is the following. During early cleavage no surface activity is evident, but at about the 64-cell stage gentle undulations begin. As cleavage continues and the segmentation cavity forms, deep blastomeres gradually form protruding, rounded, hyaline lobopodia, with increasing frequency (Fig. 2-5a). Toward the middle of the blastula stage, more elongate lobopodia appear, and both short and elongate lobopodia form rapidly and frequently every few minutes. They also withdraw rapidly, and new ones then form elsewhere on the cell surface, in turn to be soon withdrawn. During early to middle blastula these lobopodia do not adhere to other cells, and their rapid protrusion and withdrawal without adhesions cause the cells to change shape constantly but with no translocation. This "exercising in place" transforms the blastoderm into a jostling mass of surface active cells. Just before the onset of epiboly, during the late blastula stage, lobopodia begin to adhere to other deep cells and cells of the enveloping layer and can be observed to spread somewhat over their surfaces. When this occurs, the lobopodium is stretched into either an elongate filopodium or a spreading membranous fan (Fig. 2-5b). These adhesions apparently give the cells traction, for when the processes shorten the cells are pulled in the direction of the adhesion. With this, translocation of deep cells occurs. Deep cells also move without visible contraction of their cytoplasmic protrusions. But the tip of the protrusion remains the leading edge. This fact and the fact that ruffling activity can often be seen at the tip of the extended lobopodium or fan suggest that locomotory activity is localized in this region. As epiboly begins, the locomotory activity of deep cells can be observed to increase until most deep cells appear to be involved. At first, the cell movements appear to be mainly random and cause the deep cells to occupy the expanding space beneath the spreading enveloping layer. In the dorsal and lateral marginal regions, however, the movements soon acquire a directional component, and the cells tend to accumulate in the marginal region to form the germ ring and then converge dorsally to form the embryonic shield. The increase in surface adhesiveness of the deep cells toward the end of the blastula stage and during the early phases of gastrulation was predictable from the fact that they tend to flatten on glass in culture. Since deep cells crawl over each other and over the undersurface of the enveloping layer, it is clear that they are not contact inhibiting.

Surface activity during epiboly is not confined to the deep cells. Cells of the enveloping layer are also active. There is no surface displacement, however. All cells keep their positions in a cohesive layer and do not crawl over each other (Fig. 14-8). Their activities consist of contractions, expansions, and undulations. Occasionally, the membranes of two cells seem to separate over part or all of their area of contact. Cells bordering such a gap often show ruffled membrane activity or send processes across the gap. These gaps are then generally closed again after several minutes. During most epiboly, as the marginal cells of the enveloping layer crawl over the periblast, their outer, free surfaces show ruffled membrane activity. Sometimes marginal cells retract slightly, presumably because of the tension they are under. They then move toward the periblast margin again, with a renewal of surface activity (after apparently reattaching to the periblast). In contrast to the deep cells, enveloping layer cells do not move over each other and show locomotory activity only at their free margins. They therefore show contact inhibition.

The electron microscopic studies of deep blastomeres correlate closely with the time-lapse observations. Lobopodia are evident and are relatively free of organelles. This explains their hyaline appearance in the light microscope, but why they are free of organelles is unexplained. In early and middle blastulae, there are no close contacts of lobopodia with other cells. This correlates with their apparent low adhesiveness at this time.

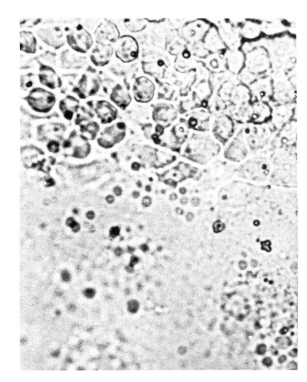

Fig. 14-8 Cells at the lateral margin of a midgastrula of *Fundulus*. Marginal cells of the flattened enveloping layer are visible at the right where they adhere to the periblast. Note that the enveloping layer cells adhere closely to each other. The margin of the periblast is located just below the margin of the enveloping layer, but it is not visible in this micrograph. ×200. (After Trinkaus and Lentz. 1967. J. Cell Biol. **32**:139.)

During late blastula and gastrula stages, however, the contact relations of deep cells change. 200-A contacts of lobopodia with other cells appear regularly (Fig. 14-9). This observation is of considerable interest. The appearance of 200-A contacts just at the time when lobopodia begin to adhere to other blastomeres supports the conclusion that such contacts are adhesive. A few close junctions have also been observed between deep blastomeres, but no desmosomes. For cells like deep cells, which are constantly in motion, the needs are special. They require some adhesion to gain traction for movement, but the adhesion must not be too strong, for otherwise the cells would become immobilized. The 200-A contacts apparently represent just this kind of adhesion.

It was not possible to visualize the relation of the lowermost deep cells to the periblast in films of living eggs, because of intervening blastomeres that obscure the view. The electron micrographs of this relation, however, are clear. They show no specialized kinds of junctions, just blastomeres lying on the tops of microvilli that protrude from the upper surface of the periblast. This relatively casual relationship is consistent with the apparent low adhesiveness of deep cells for the periblast, as revealed by

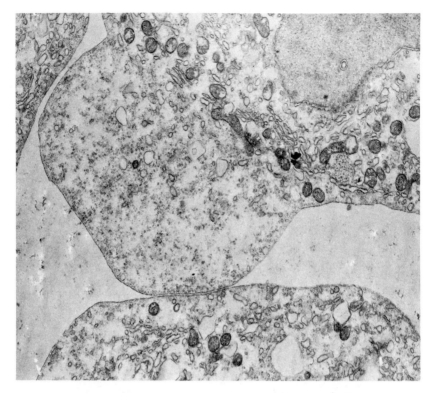

Fig. 14-9 Lobopodium-blastomere contact of deep cells at an early *Fundulus* gastrula stage. The lobopodium is relatively free of organelles, containing only a few vesicles and polyribosomes. In the zone of contact the adjacent plasma membranes are parallel and separated by a gap of 200 A. ×21,500. (After Trinkaus and Lentz. 1967. J. Cell Biol. **32:**139.)

dissections. If deep cells make use of the periblast as a substratum, they must crawl over the tips of the periblast microvilli.

Cells of the enveloping layer have quite different contact relations with each other. They are invariably bound together apically by zonal junctions (Fig. 14-10). During blastula stages this apical junction is a close junction and is the sole contact between adjacent cells. The intercellular space proximal to the junction is invariably quite wide. During epiboly, however, the junctional contacts increase. Point tight junctions are added to the close junctions in the apical zone of membrane apposition (Fig. 14-10). Proximally, the apical junction is supplemented initially by a zone in which the opposed plasma membranes are separated by a gap varying from 70–200 A. Primitive desmosomes are often present in this region. Additional close or "gap" junctions are often added to these. Also, in many places the unit membranes of adjacent enveloping layer cells follow a zig-zag course so that cytoplasmic flanges interdigitate with one another

(Fig. 14-11). It is significant that these additional junctions are added as the enveloping layer begins to be stretched by epiboly. They no doubt impede enveloping layer cells from being pulled apart during this stretching process. The presence of close junctions, tight junctions, and desmosomes

Fig. 14-10 Contact specializations between enveloping layer cells of a *Fundulus* midgastrula. The apical junctional complex (upper left) consists of both tight junctions and close junctions. The tripartite plasma membranes are in contact at the apicalmost point and at two or three other points. The condensations of electron-dense material in the cortical cytoplasm just beneath the tight junctions are unexplained. Between the tight junctions the membranes form zonal close junctions and are separated by 20–40 A. Below this apical junctional complex the opposed membranes diverge for a distance varying from 70–200 A. (Primitive desmosomes are often found in this region.) The plasma membranes then converge again to form an extensive close junction, where the membranes are separated by 20 A. This is a so-called "gap junction." It may be tight at either end. All distances are, of course, approximate. ×180,000. Inset: another apical junctional complex of a *Fundulus* midgastrula showing both tight junctions and zonal close junctions. In this instance, the plasma membranes in the close junctions are separated by 20–75 A. ×140,000. (After Lentz and Trinkaus, unpublished.)

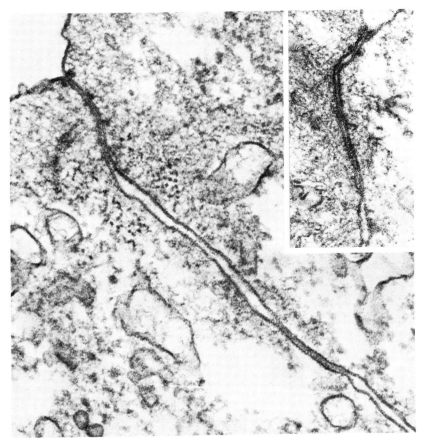

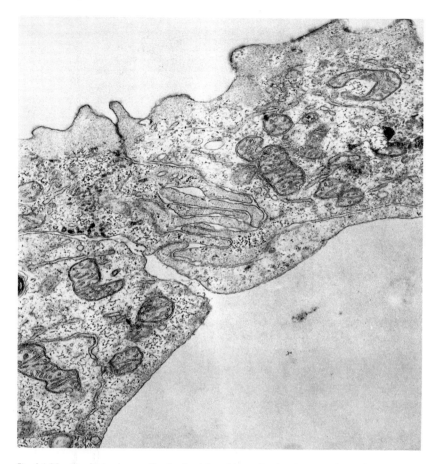

Fig. 14-11 Enveloping layer cells of a *Fundulus* midgastrula, showing several cytoplasmic flanges which interdigitate with one another. A portion of a deep blastomere (lower right) is separated from the undersurface of the enveloping layer cells by a space of 200 A. The large intercellular space in the lower right part of the micrograph is a portion of the segmentation cavity. ×25,000. (After Trinkaus and Lentz. 1967. J. Cell Biol. **32**:139.)

between tightly cohering cells of the enveloping layer sheet, as contrasted with 150–200-A gap between the highly mobile deep cells, is consistent with the general belief that these junctions are more adhesive.

In spite of the constant surface activity of enveloping layer cells, these cells remain in close contact most of the time. This is presumably due to their contact specializations. Occasionally, however, the surfaces of two cells appear in the films to pull apart and then close again. These separations are not apparent in the electron micrographs. Two possible explanations for this contradiction come to mind. The disruptions of surface contacts may occur so infrequently and close so rapidly that detection at the fine-structural level is rendered improbable. Alternatively, what appears in the films as breaks in cell contacts may be due simply to wider separation of the nonadhering parts of the opposed membranes proximal to the apical junction. Such separations could well appear as separations of whole cells at low power in a layer as transparent as the enveloping layer.

The contact relations of the marginal cells of the enveloping layer with the periblast are of particular interest, since the only firm adhesive connection of the blastoderm to the periblast is at this region. In addition, these marginal cells show ruffled membrane activity at their outer edges as they spread over the expanding periblast. Since adhesion to a substratum is necessary for cell movement, the association of these two phenomena at the edge of the enveloping layer is not unexpected. It is somewhat surprising, however, to find a close junction in this region (Fig. 14-12); for adhesions of the marginal cells of the enveloping layer to the periblast appear to be intermittent in time-lapse cinemicrography, the contacts apparently being broken and remade continuously as the cells move out over the periblast. If this is the situation, the cells must form junctions,

Fig. 14-12 Enveloping layer-periblast junction of *Fundulus* late blastula. The hyaloplasm of the periblast at the right is of lower density than that of the enveloping layer. Blunt, villous projections extend from the periblast surface. The junction between these two layers consists of a close apposition of adjacent plasma membranes. Below this junction, the membranes diverge to produce a wide intercellular space into which a few microvilli extend. A deep blastomere is at the lower left. ×22,000. (After Trinkaus and Lentz. 1967. J. Cell Biol. **32**:139.)

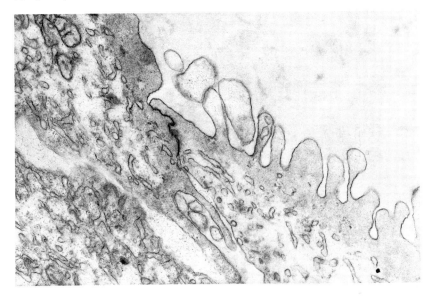

separate, and form junctions anew, in a matter of minutes. If this occurred stepwise, with a distal junction forming before the proximal contact is disrupted, there would always be at least one region of contact, as indeed is the case in the electron micrographs.

It is also possible that these close junctions remain intact most of the time, attaching the marginal cells to the periblast in a relatively permanent way. If the enveloping layer is passively dragged over the yolk by the actively spreading periblast, a rather permanent connection like this would not be unexpected. Such a situation could hold during most of epiboly, as long as a means is provided for these junctions to separate occasionally and reform in order to account for the occasional retraction and reattachment of marginal cells. The impression that the cells are adhering and deadhering intermittently in the region of their ruffled membranes could be misleading. The ruffling that one observes in the films could involve primarily the upper surface of the cell. Another possibility is that the same part of the cell surface always adheres to the periblast, sliding toward the margin of the periblast or retracting from it but never losing contact. In this case also the ruffled membrane activity would indeed involve only the nonadhering upper surface of the cell. Finally, the marginal cells of the enveloping layer could maintain continuous contact with the periblast by rolling over it. In this way, different parts of the cell surface could be brought continuously into contact with the spreading periblast without contact ever being broken.

The fine-structural studies have settled one matter that has been puzzling for years. The spherical yolk mass of teleost eggs is contained by a membrane which was termed the yolk gel layer. As we have seen, this layer has impressive contractile properties. The electron microscope reveals it to be cytoplasmic, containing mitochrondia and other organelles. It is continuous with and an extension of the periblast. This discovery modifies our view of the spreading of the periblast in epiboly. The yolk layer was thought to solate at its juncture with the periblast, while it is gradually replaced in epiboly. If now we couple the observation of continuity at the fine-structural level with a previous observation that carbon marks placed on the yolk cytoplasmic layer do not move together in epiboly (hence the yolk cytoplasmic layer does not contract), we are led to the conclusion that the periblast spreads epibolically by flowing into the yolk cytoplasmic layer, causing it to thicken and adding some organelles to it. Accordingly, periblast epiboly appears not as the spreading of a layer with abrupt margins over the noncytoplasmic surface of the yolk, but as a controlled flow of cytoplasm from the thicker animal part of an intact cytoplasmic layer (the periblast) into its thinner vegetal part (the yolk cytoplasmic layer).

It was once thought that an extraneous surface layer or "surface coat" was applied to the outer surfaces of enveloping layer cells in teleost eggs and functioned to unite them into a cohesive layer. This concept was based on insufficient evidence. The electron microscope studies of *Fundulus* development reveal no extraneous material applied to the outer surfaces of enveloping layer cells, even though the material was fixed in a way that would likely reveal the presence of a mucopolysaccharide layer. And even if such a layer is present though not revealed in the electron micrographs, it seems unnecessary. Enveloping layer cells are united by a number of junctional specializations.

Our knowledge of the junctional complexes of the blastoderm and the remarkable permeability properties of *Fundulus* eggs (they develop in both seawater and distilled water) stimulates interest in their electrical properties. Bennett and I have investigated these recently with conventional electrophysiological techniques, using glass microelectrodes. We found that cells of the enveloping layer are closely coupled electrically, in part by way of junctional contacts (presumably their apical zonal junctions) and also by way of the extracellular fluid of the segmentation cavity, which bathes their inner surfaces (Fig. 14-2). The dominance of close junctions in the apical junctional complex (Fig. 14-10) raises an interesting question. Are the areas of low resistance between enveloping layer cells confined to their tiny tight junctions or do they include the close junctions as well? If the latter is the case, the more proximal gap junctions could also be regions of low resistance. Since the enveloping layer is a contact-inhibiting system, in which the mobile activity of the cells is profoundly affected by their contact with each other, it is tempting to speculate that these influences of cells on cells may be mediated by the passage of ions across these low-resistance junctions (but see p. 208). The resistances from the cytoplasm of the cells and from the segmentation cavity to the exterior of the egg, on the other hand, are high, showing that there is an effective barrier between them and the exterior. This is provided by the external membrane of the cells of the enveloping layer and by junctions between the cells, presumably the apical zonal junctions. If it is indeed these junctions which are occluding, they would be in a functional sense true *zonulae occludentes*. These high-resistance barriers no doubt account for the extraordinary ionic independence of the *Fundulus* egg.

We are now in a position to pull together these observations and attempt to construct a comprehensive concept of the mechanism of epiboly. The enveloping layer spreads like an epithelial sheet. It is apparently contact inhibiting. Its cells do not change position relative to one another and are bound together by close and tight junctions, gap junctions, desmosomes, 200-A contacts, and interdigitations of their plasma membranes. In addition, they are electrically coupled. Moreover, and most important of all, the sheet spreads over its expanding periblast substratum as a result of the activities of its marginal cells. These form close adhesive contacts with the underlying periblast and appear to move over it by means of ruffled membrane activity at their free margins. Once the margin of the enveloping layer has reached the most marginal region of the periblast, its rate of epiboly becomes that of the periblast. It is not yet clear whether this is because the enveloping layer margin now spreads at the same rate as the periblast or whether it is being actively pulled by the spreading periblast. The strong adhesive contacts of enveloping layer cells with each other communicate the pull of the spreading marginal cells to the entire layer. Nonmarginal cells are not completely passive, however. Their surfaces are in constant activity, and when they partially separate occasionally, no doubt as a result of tension in the stretched membrane, they appear to form ruffled membranes and move together again to close the gap. Marginal cells therefore appear to be the prime movers, but nonmarginal cells are potentially active at all times and sporadically aid the spreading of the sheet by tiny spreading movements of their own. When the enveloping layer has spread entirely over the yolk, its marginal cells contact each other in closing the blastopore. As expected of a contact-inhibiting system, their spreading now ceases. During closure of the blastopore the periblast moves somewhat ahead of the enveloping layer and closes its "blastopore" first. At this time the marginal cells of the enveloping layer assume an elongate spindle shape radiating from the point of blastopore closure, suggesting an active pulling role for the marginal region. If the change in cell form is due to marginal pull, the pull must be exerted by marginal cells of the enveloping layer, since the periblast has already completed its epiboly. It is also possible that these cells become elongate not because they are stretched, but because they are compressed by each other, as they move into the ever diminishing area of the closing blastopore. They must either become longer and thinner or pile up in several layers (which they apparently do not do).

The mechanism of periblast epiboly is still not understood, even though we know that it takes place independently of the presence of the blastoderm and seems to involve the controlled flow of cytoplasm from the thicker periblast into the thinner yolk cytoplasmic layer. The observation that the periblast and yolk cytoplasmic layer are different parts of the same layer raises a puzzling question. The periblast appears to serve as a specific substratum for the blastoderm. It limits the rate of blastoderm epiboly by its own rate of spreading. But since the periblast and yolk cytoplasmic layer are simply thick and thin parts of the same layer of protoplasm, why doesn't the enveloping layer spread beyond the periblast margin? The answer to this question is not yet evident, but it may be significant that the activity and structure of the exposed outer surfaces of the two layers are markedly different. The exposed marginal surface of the periblast is highly active, with many villous projections (Fig. 14-2), whereas the outer surface of the yolk cytoplasmic layer is characteristically smooth. Perhaps the marginal cells of the enveloping layer adhere more strongly to the surface-active marginal periblast than to the yolk cytoplasmic layer.

The deep cells translocate differently from those of the enveloping layer. They are not bound together in cohesive sheets but move about over one another, over the undersurface of the enveloping layer, and probably over the underlying periblast as well. Thus they are not contact inhibiting and may be compared to the mesenchyme of echinoderm gastrulae and perhaps also to the mesoblast of the chick. At the very first they appear to move largely at random, jostling about and filling in space in the expanding segmentation cavity as epiboly begins. Soon, however, they acquire a directional component as they form the germ ring and move dorsad within the germ ring to form the embryonic shield. The deep blastomeres appear to acquire their ability to move in two stages: 1) the formation of nonadhering lobopodia during early blastula stages and 2) an increase in surface adhesiveness during late blastula and early gastrula stages. The conclusion that the adhesiveness of the cell surface increases toward the beginning of gastrulation is based on three lines of evidence: the tendency of dissociated gastrula cells to adhere and flatten on glass in cell cultures, the adhesion of lobopodia to other cells in late blastulae and early gastrulae, and the formation of 200-A junctions when lobopodia contact other cell surfaces at this stage. The lobopodium is the initial organ of locomotion, and the increase in adhesiveness gives it traction and converts it into a filopodium or fan. With contraction of the adhering filopodium or fan the cell is pulled along. The uppermost deep cells are closely applied to the under surface of the enveloping layer, and some of the lower ones are in close relation to the microvilli of the periblast. Deep cells thus move by adhering to each other, to the inner surface of the enveloping layer, and probably also to the periblast.

Even though the yolk cytoplasmic layer does not contribute to epiboly in the direct way envisioned by Lewis, its presence nevertheless is doubtless essential for the process. It exerts contractile tension tangentially and thus aids epiboly geometrically by maintaining the spherical shape of the egg and the turgor of the yolk.

In summary, we may conclude that the major forces for epiboly in *Fundulus* eggs, and probably trout eggs as well, reside in the enveloping layer and the periblast. The deep cells, which eventually form the embryo itself, do not appear to play an essential role in epiboly. Their active translocation in the expanding segmentation cavity is more concerned with the execution of morphogenetic movements which lead to embryogenesis. Since neither the enveloping layer nor the periblast engages in embryogenesis, they are in a sense uniquely morphogenetic organs. Of course, both have important permeability functions as well.

Since cells of the enveloping layer appear to be contact inhibiting and deep cells not to be contact inhibiting, their contact relations at the fine-structural level assume special importance. During epiboly, cells of the enveloping layer form close, gap, and tight junctions, 150–200-A junctions, and desmosomal junctions with each other. Also, opposing cytoplasmic flanges of adjacent cells may interdigitate. Deep cells, on the other hand, form principally only the classic 150–200-A junction with each other, very occasionally adding a close junction. The contrast in the kinds of junctions made by the two kinds of cells is so striking that this may be the basis of their contrasting locomotory behavior. It is possible that ruffled membranes of contact-inhibiting cells are paralyzed by contact with another cell surface because of their tendency to form highly adhesive contact specializations with each other. There is a need for similar studies of other contact-inhibiting and noncontact-inhibiting systems. The role of electric coupling in the contact inhibition of these cells is undetermined.

It is apparent that cells must differentiate in order for gastrulation to occur. Our correlated cinematographic, fine-structural, and cell-culture studies have demonstrated that important changes in cell surface activity take place prior to gastrulation and that these changes constitute preparation for the cell movements of gastrulation. Study of the nuclei and cytoplasm of cells during blastulation reveal differentiations here as well. During early cleavage stages the blastomeres are quite undifferentiated: the hyaloplasm is of low density, ribosomes and membranous elements of the endoplasmic reticulum are sparse, and mitochondria are relatively simple structurally. Beginning with the early blastula stage, however, the blastomeres show an increase in nuclear and cytoplasmic specializations. The nucleolus appears as a fine-textured mass of fibrillar material. Then, during the blastula stage, dense granules about 150-A in diameter and resembling cytoplasmic ribosomes appear within it. In the cytoplasm, ribosomes, particularly polyribosomes, become more numerous when the nucleolus is first noted. These changes resemble those described for the egg of *Triturus*. During the late blastula and early gastrula stages of *Fundulus* membranous elements of the endoplasmic reticulum also increase in number, and the Golgi apparatus and mitochondria become more complex. The marked increases in surface activity of the blastomeres that occur during this period presumably have their basis in these changes in cell organelles, for such differentiations involve elaboration of the synthetic machinery of the cell.

These studies of epiboly in the *Fundulus* egg have answered some questions, but inevitably they have spawned a whole family of new ones. 1) How, for example, does a filopodium or a fan of a deep cell operate as an organ of locomotion when they contract and pull the cell along? Do they terminate in ruffled membranes like those of fibroblasts, or (as appears to be the case sometimes) do they adhere over a rather broad surface to anchor the tip while the whole process is retracting? We may also ask whether the ruffled membranes of cells of the enveloping layer resemble in detail the ruffled membranes of fibroblasts. High-power microscopy of both kinds of cells under more favorable optical conditions may permit the requisite observations. 2) We also need information on how the marginal cells of the enveloping layer begin to spread. Thus far it has not been possible to obtain adequate photographs of this, because the mass of deep cells impedes observation of the margin during this stage of development. 3) If the marginal cells of the enveloping layer are indeed the prime movers in epiboly, nonmarginal cells should join the spreading progressively as their distance from the margin increases. Also, gaps between cells of the enveloping layer should appear first and more frequently near the margin, where tension is greater. 4) Although it has been shown that cells of the enveloping layer are electrically coupled and that this correlates well with their structural contacts, we as yet have no such electrical information for deep cells. 5) With increased study, the periblast has turned out to be more fascinating than ever. It would appear that the periblast spreads by a wave of surface activity accompanied by cytoplasmic flow, which progressively converts yolk cytoplasmic layer into periblast. If this is true, marks placed on the surface of an exposed periblast should remain in place (relative to the animal pole) and not move vegetally with the margin. 6) We have as yet no knowledge of the chemical events on which changes in cell behavior essential to gastrulation depend. This includes both changes in the cell surface itself and the metabolic changes that presumably lie at their basis. Now that techniques for isolating cell membranes are becoming available, it may be possible to study the membranes of *Fundulus* cells directly. Our electron microscopic studies have suggested how studies of metabolism must proceed. Techniques for this have long been available. 7) Finally, we have no understanding of the mechanism of the directionality of deep cell movements. We need careful descriptive studies of the deep cell movements of *Fundulus*, like those that have been begun for the trout. These are necessary to provide a basis for the analysis of mechanism.

SELECTED REFERENCES

Ballard, W. W. 1966. Origin of the hypoblast in *Salmo*. II. Outward movement of deep central cells. J. Exptl. Zool. **161**:211–220.

Bennett, M. V. L. and J. P. Trinkaus. 1969. Electrical coupling of embryonic cells by way of intercellular junctions and extracellular space. In preparation.

Devillers, C. 1951. Les mouvements superficiels dans la gastrulation des poissons. Arch. Anat. Microscop. Morphol. Exptl. **40:**298–309.

Devillers, C. 1961. Structural and dynamic aspects of the development of the teleostean egg. Advan. Morphogenesis. **1:**379–428.

Lentz, T. L. and J. P. Trinkaus. 1967. A fine structural study of cytodifferentiation during cleavage, blastula, and gastrula stages of *Fundulus heteroclitus.* J. Cell. Biol. **32:**121–138.

Lewis, W. L. 1949. Superficial gel layers of eggs and their role in early development. Sobretino Anales Inst. Biol. **20:**1–14.

Trinkaus, J. P. 1951. A study of the mechanism of epiboly in the egg of *Fundulus heteroclitus.* J. Exptl. Zool. **118:**269–320.

Trinkaus, J. P. 1963. The cellular basis of *Fundulus* epiboly. Adhesivity of blastula and gastrula cells in culture. Develop. Biol. **1:**513–532.

Trinkaus, J. P. and T. L. Lentz. 1967. Surface specializations of *Fundulus* cells and their relation to cell movements during gastrulation. J. Cell. Biol. **32:**139–154.

Wilson, H. V. 1891. The embryology of the sea bass (*Serranus atrarius*). Bull. U.S. Fish. Comm. **9:**209–277.

FIFTEEN

Epilogue

In this final chapter I wish to summarize some of the main points I have tried to make and discuss a few matters that are of general importance for an understanding of the contact relations between cells, but that do not fit readily into the organization of the other chapters. It will be particularly clear, I think, after reading this chapter that the specialized area of biology that is the subject of this volume has wide implications for many problems outside its immediate province.

Principles of cell motility

Although the mechanisms of most morphogenetic movements are largely obscure, certain cell properties appear to have general importance: surface adhesiveness; surface deformability; the ability to form organs of locomotion such as ruffled membranes, lobopodia, and filopodia; changes in certain structures within the cytoplasm such as microtubules and microfilaments; and contact inhibition. In some instances changes in only two or three of these may suffice to explain all; in others the situation appears to be more complex. In practically all cases, however, the really crucial detailed evidence is not yet at hand, and the literature remains largely speculative.

In particular, measurements of cell-to-cell adhesiveness and cell deformability usually have been of doubtful value or completely lacking. Concentrated efforts to accumulate information on these two surface properties are desperately needed. Fortunately, techniques are becoming available that should make the effort rewarding. The Roth-Weston technique for measuring cell-to-cell adhesiveness will certainly yield important information when applied to embryonic systems. Cell deformation could be assessed by determining the amount of negative pressure required to suck up a hemispherical bulge into a micropipette. L. Weiss has applied this technique to cells in culture and has found interesting correlations with other features of surface behavior. For example, treatment with neuraminidase, which hydrolyzes scialic acid, a constituent of the cell surface, and promotes phagocytosis, also renders the cell surface more readily deformable.

In contrast to adhesiveness and deformation, studies of cell motility, while still representing only a beginning, have shown certain clear trends. Where cells move as individuals within the organism they seem to favor the use of long, contractile lobopodia or filopodia. This is clearly the case for primary and secondary mesenchyme cells of echinoderm gastrulae and for most deep cells of the *Fundulus* blastoderm, and it probably holds as well for mesoderm cells migrating away from the streak in the chick blastoderm. Although our interest centers on modes of cell movement that contribute to early morphogenesis, it is important to emphasize that this filopodial method is not confined to cells of early embryos. Chick retinal pigment cells in heterotypic aggregates extend long filopodia, which they apparently use to pull themselves closer to other cells as they sort out. Dissociated cells from adult sponges also move about in this manner under primitive culture conditions on glass. Perhaps the most extraordinary use to which such cell processes are put has been observed in the insects. Wigglesworth has found that epidermal cells spin out filopodia up to 100 μ in length that adhere to tracheoles, and then by their contraction pull these tiny respiratory organs into regions deprived of their normal tracheal supply. The filopodia thus probably insure an even distribution of tracheoles.

When cells move as members of a cohesive sheet, their mode of movement appears different. If there is a free edge, the marginal cells appear to move in the manner of fibroblasts and epithelial cells in culture, forming ruffled membranes. A possible exception to this is the spreading chick blastoderm. The presence of very long cytoplasmic processes extending out from its marginal cells raises the possibility that in this instance the sheet is caused to spread by the contractile pull of filopodia. Another exception is the spreading of sheets lacking a free edge, as during amphibian gastrulation. The manner in which these sheets spread is a complete mystery.

Microfilaments and microtubules

Evidence that microfilaments and microtubules within the cytoplasm may have substantial importance in changing the shapes of cells during morphogenesis has already been discussed in previous chapters, but the most impressive cases remain to be described.

During metamorphosis of the ascidian *Amaroucium* the caudal epidermis contracts to a spectacular degree, diminishing to about 5% of its original length. Cloney studied the ultrastructural changes accompanying this contraction and found a remarkable change in the orientation of the microfilaments during contraction. Prior to metamorphosis the epidermal cells contain elongate and folded filaments, 50–70 Å in diameter, which are dispersed throughout the cytoplasm in a loose network. During metamorphosis many filaments become oriented in the axis of contraction and compacted in the apical cytoplasm. Significantly, filaments attach to the plasma membrane in greatest numbers at the level of maximum cellular shortening. Filament alignment and packing do not develop in noncontracting cells. During contraction, the living epidermis becomes birefringent, confirming the presence of oriented elements and indicating that the orientation of microfilaments seen in the electron microscope is not a fixation artifact. It is of course not certain whether this change in the orientation of microfilaments is a cause or an effect of the change in shape of the epidermal cells, but the correlations are so striking that at present they favor a contractile role for the microfilaments.

Further evidence that elongate intracytoplasmic structures may play an active part in changes in cell form comes from a study of spike-like protrusions from cells in culture. When first formed, these microspikes project directly into the liquid culture medium. Since they are unattached peripherally they are free to wave about in continuous activity. During this movement, however, they remain rigid, bending back and forth like rodlets hinged to the cell surface. Electron micrographs show each of these microspikes to have a central core, consisting of straight microfilaments 80 Å in diameter, and straight microtubules, 250 ± 30 Å in diameter. Since these elements are straight, they are undoubtedly rigid. Some of the microfilaments appear to be anchored to the cell membrane. These observations suggest that microspikes are projected from within the cytoplasm by microfilaments and microtubules, rather than protracted from without.

One of the most impressive studies of the possible involvement of microtubules in the shaping of cells is that of Byers and Porter on the embryonic chick lens. During their differentiation, lens cells change from a cuboidal form, about 15 μ thick, to an elongate form 3–4 μ thick and 50 μ long. It is proposed that the systems of microtubules in these cells could be an elastic "cytoskeleton" that is responsible for this enormous cellular elongation. Microtubules may lengthen by addition of subunits at their ends and thus lead to a concomitant change in cell shape. The following observations are offered in support of this thesis. a) Palisading of lens cells takes place in cultured eye rudiments that have been divested of their surrounding epidermis. This eliminates the spreading of the adjacent epidermis as a force pushing lens cells together and indicates that the mechanism of cell elongation must reside within the eye and probably within the lens cells themselves. b) The appearance of microtubules, 230 A in diameter, in the cytoplasm of lens cells takes place subsequent to lens induction and coincides with the beginning of cell elongation. c) The microtubules are oriented in the long axis of the cells, and they are straight, implying that they have the necessary rigidity. d) The microtubules disappear just before lens fiber formation commences (but they persist in those cells which continue to elongate). During this elongation process the cell nucleus also elongates somewhat and, along with mitochondria, endoplasmic reticulum, and elements of the Golgi complex, orients in the axis of cell elongation. It is not known whether the orientation of these organelles is passive, as assumed, or active, and hence a possible factor in cell elongation. The evidence in favor of the postulated causal role for the microtubules is impressive, but of course not conclusive. Moreover, there are other possible mechanisms that could also promote cell elongation, such as an increase in deformability or adhesiveness of the cell surface. These last possibilities are suggested by the fact that during cell elongation adjacent cells become more and more closely aligned. When cells are still cuboidal they are separated widely, except at the apical poles where they are tightly joined. When cells have elongated, on the other hand, the membranes of adjacent cells are in apposition over the whole area of contact. If during lens differentiation there is indeed an increase in surface adhesiveness and if the cells tended to maximize their adhesions, this could contribute to cellular elongation.

In addition to the case just described, a survey of the distribution and orientation of microtubules makes it clear that they are often associated with the intracellular migration of cytoplasmic components and with the development of pronounced asymmetries in cell shapes. For example, microtubules are an invariable accompaniment of streaming in plant cells and pigment migration in melanophores. They also form the fibers of the mitotic spindle and are found parallel to the long axis of elongated fibroblasts in developing feathers. In the light of the obvious involvement of microtubules and microfilaments in changes in cytoplasmic position and cell form, more studies are needed on the possible role of these structures during gastrulation. Are they involved in the formation and contraction of filopodia? In all of these observations an obvious question must be raised: are the microtubules (or microfilaments) the primary agents in the change in cell shape, or are they oriented passively by other as yet unknown cytoplasmic agents? Wessells and Evans, for example, failed to find a unique arrangement or quantity of microtubules in elongate cells of the developing chick pancreas. This question emphasizes the need to introduce the experimental method in all work on the causal significance of ultrastructural elements. It is known, for example, that certain antimitotic drugs, such as colchicine, inhibit the assembly or cause disassembly of the protein subunits of microtubules. This is why colchicine inhibits spindle formation and with it mitosis. Low temperature and high hydrostatic pressures also cause microtubules to break down.

Tilney has provided a particularly instructive example of the use of these treatments in his study of the role of microtubules in the formation and maintenance of the axopodial extensions of the Heliozoan, *Actinosphaerium*. Treatment causes disassembly of the microtubules, disappearance of the birefringence of axopodia, and retraction of axopodia. Within minutes after the treatment, microtubules reappear, birefringence is re-established, and the axopodia begin to re-form. In the light of these results it seems reasonable to suggest that in this instance the microtubules may not only supply by their elongation the force necessary for axopodial extension but also may give form and rigidity to each new axopodium.*

Contact inhibition

The possible involvement of contact inhibition of cell movement as a mechanism in morphogenetic movements derives from its effectiveness as a filler of cell-free space. It directs cell movements into cell-free space, and when the space is filled it stops the movement. It is self-regulatory. We should therefore look for evidence of contact inhibition where cells are at the edge of such a space. Cell sheets possessing a free edge and engaged in spreading movements would be a good example. There is already good evidence that contact inhibition is involved in the spreading of epithelial sheets in wound closure and regeneration and in the spreading of the enveloping layer in the teleost egg. Whether it is likewise involved in the spreading chick blastoderm awaits investigation. Contact inhibition has also been suggested as possibly operative in the emigration of cells from the neural crest.

It is clear now that the discovery of contact inhibition has been one of the most helpful advances in the study of cell contacts and movements. Accordingly, it is currently being subjected to intensive study in a number of laboratories. Although exploration of its mechanism has been severely handicapped by our poor understanding of the ways cells move, examina-

*An exciting possibility has been raised by some current research. Microtubules and microfilaments may not be qualitatively different; there is some indication that the filaments composing microtubules are the same as the microfilaments found free in the cytoplasm. If this is so, microfilaments may be formed in two ways: either from the breakdown of microtubules, or directly from microtubular subunits in the cytoplasm.

tion of it in more detail has revealed some provocative correlations. It now appears that the adhesion of the opposing cell surfaces of contact-inhibiting cells is represented at the fine-structural level by tight or close junctions. Conversely, noncontact-inhibiting cells, such as certain cancer cells and the deep cells of *Fundulus* blastoderms, have many fewer such junctions. These observations are of particular interest because several workers have recently shown that junctions of this sort are probably areas of low electrical resistance between cells. In other words, they are possible sites of electrical communication. In the light of this, it comes as no surprise that contact-inhibiting fibroblasts very recently also have been shown to be electrically coupled. They, and other cells possessing such junctions, are therefore able to exchange ions and probably other small molecules. The dye fluorescein, for example, has been shown to pass between coupled cells. (Some cancer cells apparently are not coupled.) It has been suggested that electrical coupling of contact-inhibiting cells is essential for the mechanism of contact inhibition, since in this manner messages from one cell could be communicated to another cell, where they could mediate the striking effects that contact-inhibiting cells have on each other (p. 175). The paralysis of the ruffled membrane, for example, no doubt involves changes in cytoplasmic viscosity and could therefore be due to the changes in ion concentration that might result from coupling with a contacted cell. The highly local nature of the paralysis of ruffled membrane activity, however, argues against the passage of such a signal transmitted through the cytoplasm. In Figure 2-3, for example, ruffling is inhibited only in that portion of the leading edge which is in contact with the other cell. Other portions of the leading edge, no more than 10–30 μ away, continue their ruffling. This result makes unlikely transmission of an inhibiting effect by diffusible molecules (such as small ions). It appears rather as if local adhesion to the surface of the other cell is what causes the ruffling to stop.

Although it is possible that this local adhesion is by means of tight or close junctions, it should be emphasized that we are not yet in a position to make a statement on this matter. The appropriate observations at the level of fine-structure have not yet been made for the type of cells illustrated in Figure 2-3 (or, for that matter, for practically all other cases where contact inhibition has been demonstrated). This lack points up a pressing need in future research on contact inhibition of cell movement for closely correlated observations on the *same cell system* with several approaches, such as time-lapse cinemicrography, electron microscopy, electrophysiology, and microinjection. Where tight or close junctions are found to be associated with both contact inhibition and electrical coupling it would seem reasonable to conclude that two probable functions of these junctions are to bind cells together adhesively and to couple them electrically. Of these two functions, the adhesive one would seem to be the more important for contact inhibition.

Although we have devoted our attention entirely to contact inhibition of cell movement, another mutual effect of cells in contact is turning out to be just as interesting. It has been established recently that when contact-inhibiting cells approach or achieve confluency in culture their mitotic activity is greatly diminished. Contact with the surface of another cell interferes in some manner with the cell cycle, and the cells rest somewhere in interphase, probably in G-1.* Not surprisingly, this inhibition of mitosis is accompanied by a progressive depression of the rates of DNA, RNA, and protein synthesis, reaching levels of 5–15% of those in a freshly inoculated culture in the case of DNA and RNA and 30–50% in the case of protein. Associated with these decreases is the disappearance of most of the free cytoplasmic polyribosomes. Upon subdivision of the culture and replating, these changes are completely and rapidly reversible. Further evidence that contact between cells can lead to changes in their metabolism comes from a study of the contact interactions between normal tissue cells and cells transformed to tumor cells by polyoma virus. These "Py cells" are characterized by a lack of contact inhibition of both movement and cell division. In consequence, their movement and multiplication in culture results in a piled up and random arrangement of cells. When such Py cells are brought into contact with normal contact-inhibiting fibroblasts, however, they are subject to inhibition of both movement and cell division. Significantly, in order for this effect to occur, the fibroblasts must be stationary in a nondividing monolayer. Stoker has studied this transfer of growth inhibition of virus-transformed cells autoradiographically and has found that incorporation of tritiated thymidine is inhibited in one- to two-thirds of the Py cells when they are in contact with stationary layers of normal cells. Py cells in the same dish that are not touching normal cells show no inhibition of thymidine incorporation. Another line of polyoma-transformed cells is deficient in the enzyme inosinic pyrophosphorylase and as a consequence fails to incorporate hypoxanthine. When these cells are placed in contact with stationary normal cells they too reacquire normal activity and, in this case, incorporate hypoxanthine. These observations indicate that certain substances can pass directly from normal to transformed cells.

Although the fine-structural, electrical, and chemical bases of these cell interactions are just beginning to be explored, it is already evident that the implications are great, not only insofar as they concern the interactions between individual cells, but for the embryo as a whole. Interactions of the sort now being studied may well lie at the basis of much of the

*In this connection it is interesting that when myoblasts fuse to form the syncytial myofibril their mitotic activity ceases completely, and there is no incorporation of tritiated thymidine. The resemblance to mitotic inhibition by contact with other cells is striking, and it seems unlikely that this is coincidental. Common causes may well be at play in both cases.

coordinated collective activities of cells and cell sheets during early morphogenesis and during subsequent tissue and organ formation.

Contact inhibition has obviously emerged as a phenomenon whose further study seems likely to reveal as much about the in vivo activities of cells as about their behavior under the artificial conditions of tissue culture. The study of the movements of cells in mixed cell aggregates, on the contrary, in spite of the attention it has attracted, seems now less likely to teach us as much as we have thought it would about in vivo phenomena. Locomotion of cells in culture and during morphogenetic movements in the embryo shears the cells across whatever substratum it is adhering to and, as we know, can result in cells translocating relatively enormous distances. Associative movement within a mixed cell aggregate, on the contrary, if we can judge from the movements of retinal pigment cells, is a movement that draws a cell toward a cell that it is already adhering to and consequently moves it only very short distances. In addition, whereas contact inhibition may be studied under the close-to-ideal conditions of cell culture, sorting out in mixed aggregates is largely hidden from the eye and hence from much detailed observation.

The cell membrane as an autonomous unit

It is apparent from studies of contact inhibition and cell associations in mixed aggregates and in the embryo that the contact behavior of different kinds of cells may differ in characteristic ways. There is much reason to believe that the decisive control of this behavior lies in the properties of the cell membrane. In many instances this is a stable cell property. The membrane of a particular cell type may vary in its contact behavior at certain times, such as during the cell cycle or during the early phases of sorting out in a mixed aggregate, but it always returns to its characteristic state afterward. The cell membrane is thus a persistent cell organelle. It does not disappear or disorganize as the cell divides or grows, but remains intact, growing as the cell grows and doubling in amount by the time a cell divides. A differentiated cell may thus retain certain membrane properties during many cell generations. The process whereby a cell maintains the stability of its membrane while growing is not at all well understood, but, because the membrane remains intact, its growth appears to involve the interpolation of new molecules into the existing membrane structure. How a cell changes its membrane properties, as we know it does during the course of morphogenesis or during the differentiation of neoplastic cells, is likewise poorly understood. This differentiation from one stable state to another also presumably involves changes in membrane structure.

There appear to be two possible sources of these structural changes. First, if the membranes of different cells are characterized by different kinds of molecules, differential gene function with reference to membrane molecules would be essential for the differentiation of new membrane material. This possibility cannot be tested until we are able to determine the detailed composition of the membranes of different kinds of differentiated cells and of cells in various stages of their differentiation. A second possible source of changes in membrane structure is simply a change in the proportions of existing membrane molecules or in their relationships to one another. For this type of change the involvement of new gene activity would not be necessary.

This second possibility is worth considering seriously, for Sonneborn has presented evidence that an alteration of the cell membrane of *Paramecium* can be reproduced in the complete absence of genic change. When pieces of the cortex of one *Paramecium* were grafted onto another, a modified cortical pattern was produced and maintained through as many as 700 fissions; that is, the membrane was permanently altered. Breeding experiments showed that there were no genic changes during this period. From this it was concluded that the "presence, location, and shape of newly formed structures is determined by the cortical environment existing at the time of their development." That the altered differentiation was not due to the failure of genes to produce the requisite molecules was shown by the coexistance of normal and abnormal regions on the same cell. Nor were the changes in membrane structure due to persisting differences in local concentrations of molecules; the regions of altered cortical pattern persisted as they moved around the cell in successive cell generations. The conclusion is forced on us that the differentiation of new membrane with a particular pattern in *Paramecium* can be autonomous from the genome and determined by the pattern of preexisting units in the membrane itself, acting as a template for the assembly of additional units to make more membrane of the same structure. This idea presupposes that the same kinds of molecules can be arranged in different ways.

These investigations establish beyond question that the organization of the cortex or membrane of *Paramecium* is a self-replicating system, perpetuating itself independently of gene activity. The genes, of course, provide the molecules out of which the membrane is formed, but the arrangement of these molecules is controlled by the preexisting arrangement of molecules within the membrane, acting as templates.

Whether or not the membranes of metazoan cells are similarly autonomous is largely a moot question, but there is evidence that the cortex of the egg (plasma membrane plus immediately underlying cytoplasm) is a highly stable system with basic effects on development. In the Ctenophore, *Beroe,* cortical material is birefringent in the one-cell stage, indicating a considerable degree of molecular orientation, and may be readily traced through all the early cleavage divisions until it is segregated entirely in the micromeres, which later form the ciliated bands of the larva. In the mosaic molluscan egg there is some suggestion that the cortex controls the localization of various biochemically differentiated types of cytoplasm shortly after fertilization and that these in turn perform the differentiations necessary to produce the various cell types in their correct positions. In the insects the cortex is involved with the activation of the nuclei into their varying roles in differentiation and with the extraordinary chromosomal eliminations that occur in certain insect species.

In the Amphibia, the well-known gray crescent is a striking cortical differentiation in the fertilized egg that preserves its integrity throughout cleavage and blastulation to determine later the locus of blastopore formation. The experiments that clinched its morphogenetic role were performed by Curtis and consisted of grafting pieces of gray crescent cortical material from fertilized eggs of *Xenopus* to the ventral regions of other eggs. Such grafts resulted in the appearance of a secondary dorsal lip and a subsequent secondary axis. Transplantation of underlying yolk cytoplasm has no effect. If the gray crescent is completely excised, cleavage and mitosis continue unimpaired, but there is no development of embryonic structures at all. Other regions of the cortex seem unimportant in the initiation of morphogenesis.

Although these experiments demonstrate that cortical material of the gray crescent remains distinct from the rest of the cortex throughout cleavage and blastulation, they do not indicate whether the cortex of the *Xenopus* egg possesses a self-replicating stability that approaches that of *Paramecium.* Curtis attempted to gather information on this matter by injuring the gray crescent mechanically in a series of eggs and following the effects in succeeding generations. The effect, an arrest of development just before gastrulation, appeared at a low frequency in the first generation, but increased in frequency in the two succeeding generations until in the third generation 85% of all cleaving eggs from experimental females showed pregastrula arrest, in comparison with 0.5% of the eggs from control females. Some experimental females gave rise to embryos that showed 100% arrest. These results show that disturbing a part of the cortex essential for its correct replication results either in failure to replicate correct cortical properties in subsequent generations or in establishing a new population of replicating molecules responsible for the effect on gastrulation. No decision can be made at present between these two possibilities, nor is it known whether or not the cortex is autonomous of the genome, but the results clearly imply that the gray crescent cortex could have self-replicating properties and that the process of replication is highly stable.

If the structure of the cortex of the *Xenopus* egg and of other metazoan cell membranes is determined by the template activity of preexisting membranes, then some environmentally imposed variation in membrane structure could lead directly to a stable heritable change in contact properties, such as a change in adhesiveness. This could occur in normal development as a result of inductive activity of adjacent cells. It also could occur in the transformation of normal tissue cells into malignant neoplastic cells by carcinogens, viruses, or other unknown agents. The point is that the striking changes in the cell membrane that take place in normal development and during carcinogenesis *may* not involve gene activation but only changes in the membrane itself and *may* be transmitted to succeeding cell generations by the self-replication of the membrane, without the intervention of nuclear genes.

Genic control of morphogenetic movements

Even though some changes in cell membranes may be dependent directly on the environment, others may depend on nuclear genes. Various interspecific hybrids between the eggs of *Rana pipiens* and the sperm of other species of *Rana* provide a case in point. These hybrid embryos develop normally until the onset of gastrulation, at which time they cease development. Nuclear genes coming from the foreign sperm are unquestionably involved in this developmental arrest, and they appear to accomplish it by interfering with the morphogenetic movements of gastrulation. Since we have ample evidence that morphogenetic movements depend on changes in cell contact and motile behavior, i.e., primarily on changes in cell membranes, it is logical to assume that the genes in these hybrids act in some way to prevent the normal changes in membrane properties that are fundamental to gastrulation (for evidence, see p. 156).

There is a double lesson to be drawn from these results and this line of reasoning. In the first place, nuclear genes may well be active, in some instances, in controlling changes in membrane properties. In the second place, in order to understand how gene transcription and translation relate to gastrulation (and other morphogenetic movements) we must know the discrete cellular changes involved. It is meaningless to talk about genes "controlling gastrulation." It is meaningful only to discuss genic control of detailed cellular events, and eventually their biochemical basis. Genes could affect a morphogenetic movement, for example, by controlling the synthesis of proteins important in the structure of cell membranes; for this in turn could affect membrane adhesiveness or deformability. It is apparent that as we become able to catalogue the discrete cellular changes that are at the basis of morphogenetic cell movements and the changes in the molecular structure of cell membranes or other organelles that are at the basis of these cellular changes, we will be in a position to study the mechanisms of genic control of morphogenesis at a molecular level. This will enable us for the first time to forge some of the links that connect morphology and molecular biology.

SELECTED REFERENCES

Beisson, J. and T. M. Sonneborn. 1965. Cytoplasmic inheritance of the organization of the cell cortex in *Paramecium aurelia*. Proc. Natl. Acad. Sci. **53**:275–282.

Bennett, M. V. L., G. D. Pappas, M. Giménez and Y. Nakajima. 1967. Physiology and ultrastructure of electrotonic junctions. IV. Medullary electromotor nuclei in gymnotid fish. J. Neurophysiol. **30**:236–300.

Byers, B. and K. Porter. 1964. Oriented microtubules in elongating cells of the developing lens rudiment after induction. Proc. Natl. Acad. Sci. **52**:1091–1099.

Cloney, R. A. 1966. Cytoplasmic filaments and cell movements: epidermal cells during ascidian metamorphosis. J. Ultrastruct. Res. **14**:300–328.

Curtis, A. S. G. 1965. Cortical inheritance in the amphibian *Xenopus laevis*: preliminary results. Arch. Biol. (Leige). **76**:523–546.

Furshpan, E. J. and D. E. Potter. 1968. Low-resistance junctions between cells in embryos and tissue culture, p. 95–127. *In* A. A. Moscona and A. Monroy [ed.] Current topics in developmental biology, vol. III. Academic Press, New York.

Green, H. and G. J. Todaro. 1967. The mammalian cell as differentiated microorganism. Ann. Rev. Microbiol. **21**:573–600.

Levine, E. M., Y. Becker, C. W. Boone and H. Eagle. 1965. Contact inhibition, macromolecular synthesis, and polyribosomes in cultured human diploid fibroblasts. Proc. Natl. Acad. Sci. **53**:350–356.

Loewenstein, W. R. 1966. Permeability of membrane junctions. Ann. N.Y. Acad. Sci. **137**:441–472.

Martinez-Palomo, A., C. Brailowsky and W. Berhard. 1969. Ultrastructural modification of the cell surface and intercellular contacts of some transformed cells. Canc. Res. **29**:925–937.

Rubin, H. 1966. Fact and theory about the surface in carcinogenesis, p. 315–337. *In* M. Locke [ed.] Major problems in developmental biology. 25th Symposium of Soc. Devel. Biol.

Sachs, L. 1967. An analysis of the mechanism of neoplastic cell transformation by polyoma virus, hydrocarbons, and x-irradiation, p. 129–150. *In* A. A. Moscona and A. Monroy [ed.] Current topics in developmental biology, vol. III. Academic Press, New York.

Stoker, M. G. P. 1967. Transfer of growth inhibition between normal and virus-transformed cells: autoradiographic studies using marked cells. J. Cell Sci. **2:**293–304.

Taylor, A. C. 1966. Microtubules in the microspikes and cortical cytoplasm of isolated cells. J. Cell Biol. **28:**155–168.

Tilney, L. G. and K. R. Porter. 1967. Studies on the microtubules in Heliozoa. II. The effect of low temperature on these structures in the formation and maintenance of the axopodia. J. Cell Biol. **34:**327–343.

Wessells, N. K. and J. Evans. 1968. The ultrastructure of oriented cells and extracellular materials between developing feathers. Develop. Biol. **18:**42–61.

Wigglesworth, V. B. 1959. The role of the epidermal cells in the migration of tracheoles in *Rodnius prolixus* (Hemiptera). J. Exptl. Biol. **36:**632–640.

Author Index

Abercrombie, M., 20, 24, 25, 34, 35, 129, 130, 170
Allen, R. D., xvi, 16, 35
Ambrose, E. J., 20, 21, 24, 34, 35, 83, 85, 89, 98
Anderson, P. N., 52
Armstrong, P. B., 181
Arora, H. L., 105

Baker, P. A., 148, 158, 161, 163, 165
Balinsky, B. I., 15, 158, 172, 177
Ballard, W. W., 11, 15, 179, 201
Bangham, A. D., 86, 98
Barkley, D. S., 53
Becker, Y., 214
Beisson, J., 214
Bellairs, R., 166, 168, 177
Benedetti, E. L., 98
Bennett, M. V. L., 197, 201, 214
Berman, I., 15
Bernhard, W., 214
Betchaku, T., 23
Bonner, J. T., xvi, 16, 45, 46, 47, 48, 50, 52, 53
Boone, C. W., 214

Boyden, J. T., 42, 53
Brailowsky, C., 214
Brock, T. D., 78, 98
Brokaw, C. L., 38, 53
Brown, M. G., 165
Burt, A. S., 136, 139, 145
Butler, E. G., 13
Byers, B., 205, 214

Carter, S. B., 25, 28, 31, 35, 37, 123, 276
Chang, Y. Y., 53
Chiakulas, J. J., 103, 124
Child, S. W., 181
Cloney, R. A., 144, 205, 214
Colwin, A. L., 64–67
Colwin, L. H., 64–67
Coman, D. R., 27, 68, 69, 74
Crandall, A. M., 98
Curtis, A. S. G., x, xvi, 67, 72, 74, 80, 84, 85, 93, 98, 110, 113, 114, 115, 122, 124, 212, 214

Dan, K., 32, 83, 138, 140, 141, 145
Danielli, J. D., 56
Davis, B. D., 98

Author Index

DeHaan, R. L., 130, 175–177
Devillers, C., 181, 182, 185, 186, 202
DuBois, G., 14, 16

Eagle, H., 214
Easty, G. C., 83, 98
Eliot, T. S., ix
Elsdale, T. R., 124, 156, 158
Emmelot, P., 98
Ephrussi, B., 67
Evans, J., 206, 215

Farquhar, M. G., 56, 67
Fawcett, D. W., xvi, 57, 67
Flickinger, R. A., 14
Forrester, J. A., 89
Fukushi, T., 145
Furshpan, E. J., 214

Galtsoff, P. S., 101, 125, 129
Gerisch, G., 47, 53
Gibbons, J. R., 144, 145
Gillette, R., 162, 165
Giménez, M., 214
Gitlin, G., 170
Giudice, G., 98
Glaser, O. C., 162
Goldacre, R. J., 19, 35
Granholm, N., 172
Green, H., 214
Gregg, J. H., 77
Grobstein, C., 93, 94, 98
Gross, M., 9
Groves, P., 104, 106
Gustafson, T., xvi, 12, 32, 33, 35, 137, 138, 140–145, 163

Haas, H., 169
Hamburger, V., 158, 165
Harris, A. K., 22
Harris, H., 53
Harrison, R. G., ix, 28, 35
Harvey, E. B., 56
Hay, E. D., 61, 171–173, 175, 177, 178
Heaysman, J. E. M., 35
Herbst, C., 79, 98
Hilfer, S. R., 125
His, W., 159, 162
Holmes, S. J., 130
Holtfreter, J., ix, 102–104, 107, 109, 114, 115, 125, 129, 134, 147, 149–151, 155, 156, 158, 164, 165

Holtzer, H., 67
Hooker, D., 104
Hörstadius, S., 16, 145
Howze, G. B., 99
Humphreys, T., 96, 98
Huxley, J. S., 101

Ingram, V. M., 22

Jacobson, C., 160, 165
Jacoby, F., 16
Johnson, K. E., 156, 158
Jones, K. W., 85, 124, 156, 158

Kamiya, N., xvi
Kessell, R. G., 189
Kinnander, H., 138, 145
Konijn, T. M. D., 53
Krulikowski, L. S., 23
Kuroda, Y., 124

Lachmann, P. J., 75
Lash, J. W., 130
Leblond, C. P., 63, 67
Lehmann, F. E., 152
Lentz, J. P., 119, 121, 125
Lentz, T. L., 43, 181, 191–195, 202
Lesseps, R. J., 67, 86, 87, 98
Levine, E. M., 214
Lewis, W. H., ix, 109, 130, 163, 165, 181, 183, 199, 202
Lilien, J. E., 95, 98
Lillie, F. R., 133, 134
Loewenstein, W. R., 214

Mangold, H., 152
Martinez-Palomo, A., 214
Mercer, E. H., 67
Meyer, D. B., 6, 16
Middleton, C. A., 130
Miller, R. L., 44, 54
Millinog, G., 98
Mintz, B., 7, 16, 67
Monahan, D., 112
Moore, A. R., 136, 139, 141, 145
Moscona, A. A., 94, 95, 96, 99, 102, 104, 106, 117, 124, 125

Nakajima, Y., 214
New, D. A. T., 166, 167, 177
Nicholas, J. S., 11
Niu, M. C., 41, 54

O'Brien, J. P., 13
Okada, T. S., 108, 124, 125
Okazaki, K., 32, 138, 140, 141, 145
Oldfield, F., 54
O'Neill, C. H., 19, 35
Overton, J., 60, 177

Palade, G. E., 56, 67
Pappas, G. D., 214
Perry, M. M., 158, 163
Pethica, B. A., 84–87, 98, 99
Pfeffer, W., 38
Picken, L., xvi
Poole, J. P., 99
Porter, K. R., 144, 205, 214, 215
Potter, D. E., 214

Rambourg, A., 63, 67
Raper, K. B., 47
Rappaport, C., 81, 82, 99
Revel, J. P., 61, 172, 175, 178
Rhumbler, L., 149, 151, 158, 162
Richart, R. M., 131
Robertson, J. D., 67, 85, 92
Rosen, W. G., 54
Rosenberg, M. D., 30, 35, 93, 95
Rosenquist, G. C., 171, 177
Roth, S., 73, 74, 116, 121, 203
Rothschild, L., 38, 39, 54
Roux, W., 40
Rubin, H., 214
Rudnick, D., 177
Ruffini, A., 149

Sachs, L., 215
Samuel, E., 47, 49
Schechtman, A. M., 11, 16, 152, 158
Schlesinger, A. G., 169, 177
Schmitt, F. O., 165
Schroeder, T. E., 161, 163, 165
Selman, G. G., 164, 165
Shaffer, B. M., 47–52, 54
Simon, D., 5, 6, 16
Sonneborn, R. M., 211, 214
Spemann, H., 152
Sperry, R., 104, 105, 125
Spiegel, M., 77, 78
Spratt, N. T., 166, 169, 177

Stableford, L. T., 151, 156, 158
Stefanelli, A., 109
Steinberg, M. S., 15, 70, 71, 74, 95, 110, 111, 113–116, 122, 151, 155, 156, 158
Stockdale, F. E., 67
Stoker, M., 209, 215
Sussman, M., 47, 52, 54

Taylor, A. C., 82, 95, 123, 144, 215
Taylor, G. I., 130
Thompson, C. M., 34
Tilney, L. G., 144, 145, 207, 215
Todaro, G. J., 214
Townes, P. L., 104, 107, 115, 125
Trelstad, R. L., 172, 175, 178
Trifonowa, A., 134
Trinkaus, J. P., xvi, 9, 14, 16, 23, 34, 91, 104, 106, 112, 117, 119, 121, 125, 127, 128, 131, 138, 140, 143, 181, 182, 185, 187, 191–195, 197, 201, 202
Twitty, V. C., 41, 54
Tyler, A., 77, 99

Vakaet, L., 170, 178
Vaughan, R. B., 127, 128, 131
Vogt, W., 3, 16, 18, 149, 152

Waddington, C. H., xvi, 149, 157, 158, 163, 164, 166, 178
Walther, H. H., 172, 177
Warren, L., 98
Weiss, L., x, xvi, 16, 69, 72, 75, 80, 97, 99, 116, 204
Weiss, M. C., 67
Weiss, P., ix, xvi, 22, 28–31, 35, 70, 77, 78, 93, 95, 99, 103, 123
Wessells, N. K., xvi, 206, 215
Weston, J., 8, 10, 16, 73, 74, 116, 121, 203
Wigglesworth, V. B., 204, 215
Wilbanks, G. D., 131
Willis, R. A., 16
Wilson, H. V., 100, 101, 202
Wilt, F. H., xvi
Wolff, E., 14, 15
Wolpert, L., xvi, 12, 19, 33, 35, 65, 137, 138, 140, 141, 143, 144, 145, 163
Wood, R. L., 67
Wright, E. E., 52

Subject Index

Acrasin:
  cellular slime molds, 48–53
    3′,5′-adenosine monophosphate, 53
    effect on amoeboid movement, 48
    production, specificity, chemical nature, 52, 53
    production in cell streams, 49, 50
    secretion fronts, 51
Actinosphaerium, microtubules in axopodia, 207
Active cell movement:
  enveloping layer teleost gastrulation, 198
  spreading cell sheet, 133
    chick blastoderm, 168
    epithelial cell sheet, 127, 128
3′,5′-Adenosine monophosphate (*see* Acrasin)
Adenosine triphosphate:
  amoeboid movement, 19
  gliding movement, 22
Adhesion, cell, 18, 19, 26–28, 31, 48, 51, 55–130, 155, 156, 176, 188, 200, 203, 204 (*see also* Cell adhesion)
Adhesion, differential, 6, 9, 27, 28, 30, 110–

Adhesion (*cont.*):
  112, 115, 116, 118, 124, 144 (*see also* Differential cellular adhesion)
Adrenal medulla, formation, 7
Agglutination:
  cellular slime mold, 77
  erythrocytes, 70, 78
  factor, 78
  microvilli of dissociated cells, 86
  yeast, mating types, 78
Aggregation, 45, 47–52, 72, 79, 87, 95, 100–124 (*see also* Cell aggregation)
Alkaline phosphates, mouse germ cells, 7
Amaroucium, caudal epidermal contraction, 205
Amblystoma maculatum, neurulation, 162, 165 (ref.)
Amide bonds, mechanism cell adhesion, 76
Amine:
  bonds, mechanism cell adhesion, 76
  groups, mechanism cell adhesion, 77
Amoebae (*see* Cellular slime molds)
Amoeba proteus, amoeboid movement, 35 (ref.)
Amoebocytes, migratory powers, 13

Subject Index

Amoeboid archaeocytes, aggregation, 101
Amoeboid movement:
  cellular slime molds, 35, 45–53 (see also Cellular slime molds)
  ectoplasm, 19
  electron microscopic study, 18
  fluorescein marking, 19
  leukocytes in vivo, 19, 32, 36
  lymphocytes in vivo, 32, 42
  plasma sol and gel, 18
  reaction to antibodies, 19
  theories, 18–20
AMP (see 3′,5′-Adenosine monophosphate)
Amphibian embryonic cells:
  adhesive and aggregative properties, 124
  cell cluster movement, 129
  mixed cell aggregates, 104, 106
  presumptive epidermis in vitro, 102
Amphibian gastrulation, 4, 13
  archenteron formation, 3, 152, 156, 157
  blastopore, 11, 154
    formation and deepening, 147–150
  blastula:
    adhesive and aggregative properties, 124
    dissociation effect of calcium-free media, 156
    relation to bottle cell formation, 149
  bottle cells, 11, 147–149, 151, 155, 164
    (see also Bottle cells)
  cell adhesiveness, 155, 156
  cellularization, 133, 134
  desmosomes, cell surface layer, 157
  dissociation effect of calcium ions, 156
  ectoderm, 3, 102, 104
    spreading, 154–156, 158
  endodermal spreading, 3, 11, 149, 152–157
  germ layers:
    cellular segregation, 115
    distinguishing characteristics, 104
    fine structure, 157
    in vitro, 146
    role of calcium in cohesion, 156
    selective adhesion, 102
  invagination, 2, 17, 18
    comparison with neural closure forces, 164
  forces involved, 151, 152, 157
  formation and function of bottle cells, 147–151
  similarity with echinoderm, 151

Amphibian gastrulation (cont.):
  simulation in culture, 150
  involution:
    neck cell movement, 149
    outer cell layer, 147
  mesoderm, 11, 133, 152, 155, 156
  microtubules, 157
  neck cells, 148–150 (see also Bottle cells)
  neurula, 124, 152, 154, 156
  pH gradient, 149, 151
  reaggregation and recovery time, 114
  reconstitution of pronephric system, 102
  spreading of cell sheets, 204
  surface coat, 146–147
  theoretical explanation, 17
Amphibian gray crescent, cortical differentiation, 211, 212
Amphibian neurulation, 3, 159–161, 163
  (see also Neurulation)
Amphibian regeneration, blastema formation, 13, 14
Amphioxus, secondary phase invagination, 142
Anaplastic cells, 22
Anastomosis, cell migration, 5
Annelid:
  activity in cell sheets, 133, 134
  plasma membrane, 57
  sperm and egg fusion, 64, 65
Antibodies:
  cell adhesion, 77, 78
  effect on amoeboid movement, 19
  flourescein-labeled, 70, 108, 124
Antigen-antibody complexes, chemotactic agents, 43
Antigenic material:
  agglutination of erythrocytes, 70
  cell adhesion, 70
Antigens, type-specific cell adhesion, 77
Aorta:
  accumulation of germ cells, 6
  neural crest cells, 10
Aphrodite, plasma membrane, 57
Arbacia punctulata:
  microtubule formation in filopodia, 145 (ref.)
  septate desmosomes, 61
Archaeocytes (see Amoeboid archaeocytes)
Archegonium, 40
  pores, 27
Archenteron:
  elongation in sea urchin, 142

Archenteron (*cont.*):
  endodermal spreading in amphibian, 154
  formation:
    amphibian, 3, 152, 156, 157
    echinoderm, 12, 32, 33, 136–138
  formation of neural tube, 159
  invagination during endoderm gastrulation, 133
  relation to bottle cells, 148
  relation to secondary mesenchyme, echinoderm, 12, 141–143
  wall, relation to outer cell layer, amphibian, 147
Ascidian tadpole, contractile system, 144, 205
ATP (*see* Adenosine triphosphate)
Attachment cone, formation in sea urchin, 142
Auditory vesicles, formation, 4
Axolotyl:
  measurement of neural closure forces, 164
  neurulation, (fig.) 160, 161
Axopodia of *Actinosphaerium*, 207

Bacteria:
  chemotactic agent, 40, 42
  phage attachment, 86
  social phase in slime molds, 47
Basal lamina in chick mesodermal contacts, 173
Basal layer in chick epiblast, 172
Beroe, cortex in development, 211, 212
Blastema:
  amphibian regeneration, 13, 14
  planaria regeneration, 14, 15
Blastoderm:
  definition, 3
  relation to periblast in meroblastic egg, 66
Blastomere:
  cytoplasmic bridges, 64
  migration, 3, 11
Blastulation, definition, 11
Blood chemistry, calcium binding, 81
Blue-green algae, gliding movements, 20
Bone, intercellular composition, 90
Bottle cells:
  amphibian:
    electron microscopic study, 148
    formation, 147, 148, 156, 157
    function, 149–151
    microfilaments, 151, 157

Bottle cells (*cont.*):
    microtubules, 157, 164
    microvilli, 148
    migration, 11
    mouth opening, formation, 149
    nasal placode formation, 149
    neck cells, 148–150
    relation to archenteron surface, 148
    relation to contracting endoderm, 155
  chick, microtubules, 172
Brachydanio rerio (*see* Zebra fish)
Bracken fern, sperm movement, 36–38
Brain formation, 4
Brownian movement, relation to cell adhesion, 84, 86

Calcium, relation to cell adhesion, 156
Calcium binding, cell adhesion theory, 79–81
Calcium-free media, effect on dissociation, 19, 79, 81, 90, 156
Campanularia, chemotaxis in sperm migration, 43–45, 54
Cancer cells:
  absence of contact inhibition, 22, 24, 25, 68, 207
  adhesion, 13, 15, 22, 26, 27
  adhesive forces, 68
  anaplastic cells, 22
  cell-to-cell adhesion, 109
  electrical coupling, 208
  electronegative charge, 83
  formation by polyoma virus, 89
  glycolysis, 25
  migration of neoblasts, 13, 15
  neoplastic, measurement adhesive forces, 68
  sarcoma cell movements, 22, 24, 25
Capillarity, chemotaxis, 38–41
Carbon marking:
  chick gastrulation, inadequacies, 8
  hypoblast movements, 169
  teleost gastrulation:
    blastoderm spreading, 185
    yolk cytoplasmic layer, 196
Carboxyl groups, calcium binding sites, 79
Cardiac propulsion in embryo, 6
Cartilage:
  cellular segregation, 103, 104, 106, 107
  intercellular composition, 90
  visceral, 7

Cell adhesion:
  aggregates, 32, 50, 55, 86, 87, 100–109, 116–121, 162, 203
  amide and amine bonds, 76, 77
  amoeboid movement, 18, 19
  antibodies, 77, 78
  antigenic material, 70
  antigens, 77
  behavior, explants, 24, 46
  Brownian movement, 84, 86
  calcium-bridge hypothesis, 79–81, 83, 156
  cancer cells, 13, 15, 22–27, 68, 83, 89, 109, 207
  cell-to-cell, 72–74 (see also Cell junctions)
    embryonic spinal cord, 104
    measurement, 203, 204
    relation to intercellular material, 90–97
    teleost epiboly, 191–196
  cellular slime molds (see also Cellular slime molds):
    center formation theory, 48
    secretion front theory, 51
  chemical linkages, 76, 77
  cohesive strength, 70
  contact guidance, 28–31
  contact inhibition, 22–26 (see also Contact inhibition)
  cytoplasmic bridge, 63, 64
  degree of flattening, 69
  differential (see Differential cellular adhesion)
  distraction time and force, 70–72
  effect of high cell density, 26
  effect of temperature change, 72
  electrostatic forces, 83–86
  ester-linkage, 76
  fine structure, 55–65 (see also Cell junctions)
  forces, 55, 83–86
  fusion, 63–66, 209
  glass substratum, 26–31, 70, 82, 83, 95, 96
    role of fetuin, 83
  gliding movement, 20–22
  heterotypic, 73, 111, 112, 116, 118
  hydrogen bonding, 76
  intercellular material, 74, 90–97
  interdigitation, 62
  intermolecular bonding, 90
  in vivo cell movements (see specific morphogenetic movement)
  ion pairing, 77

Cell adhesion (cont.):
  isotypic, 73, 111, 116, 118
  magnesium, 79
  measurement, 68–74
  mechanisms, 59, 76–98
  mixed aggregates, 69, 86 (see also Reconstitution)
  molecular complementarity hypothesis, 76–79
  neoplastic cells, 69
  neurulation, 159–165
  nonspecific, 104
  organs (see Cell locomotion)
  palladium, 25, 28, 31, 37
  phospholipids, 85
  physical separation, 68–72
  physical theories, 83
  physiochemical nature, 31
  potassium ion effect, 81, 82
  proteins, 83, 95, 96, 213
  quantitative differences, 116
  reaggregation kinetics, 68, 72–74
  reduction during maturity, 27
  selective, 31, 55, 102–104
  selective deadhesion, 97–98
  sensitivity to pH, 24
  strontium, 79
  substratum specificity, 26–31, 69–72
  syncytium formation, 65, 66, 209
  thiol linkage, 76
  van der Waals-London forces, 84, 85
Cell aggregates (see also Cell aggregation):
  cell surface properties, 162
  chick retinal pigment, 203
  formation, 50
  mixed (see also Reconstitution):
    adhesion, 56, 69, 86
    cell origin, 104, 106, 107
    cellular segregation, 116–121
    cluster enlargement, 117–121
    reconstitution, 100–109
    tool for study of cell movements, 210
  selective adhesion, 32, 55
Cell aggregation, 100–124 (see also Cellular segregation):
  amphibian gastrulae, 114, 124
  cellular slime molds:
    acrasin, 48–53 (see also Acrasin)
    cell streams, 50, 51
    effects of steroids, 52
    founder cell, 48

Cell aggregation (*cont.*):
  initiator cells, 45, 48
  chemotactic factor, 95, 109, 110 (*see also* Acrasin)
  chemotactic factor, 95, 109, 110
  chick cells, 72 (*see also* Chick embryonic cells)
  interference, 78
  reconstitution, 100–109
  sea urchin mesenchyme blastulae, 87
  surface changes during dissociation, 110
  time-lapse cinemicrography, 79
Cell cement (*see* Intercellular, material)
Cell contacts (*see* Cell junctions)
Cell deformability, measurement, 203, 204, 213
Cell fusion, 63
  cell cultures, 65, 66
  definition, 64
  junctions (*see* Cell junctions)
  myoblasts, 64–66, 209
  nonmuscle, comparison with muscle, 66
  sperm and egg, 64, 65
  syncytium formation, 64, 66, 209
Cell junctions:
  close:
    cell adhesion during amphibian epiboly, 157
    chick ectodermal cells, 169
    coupling, electrical, 58
    deep blastomeres of teleost, 191
    definition, 58
    enveloping layer of teleost, 192, 193, 197, 200
    isolated cell membranes, 89
    marginal cells of enveloping layer and periblast in teleost, 196
    mesodermal cell contact in chick, 173
    opposing cell surfaces of contact-inhibiting cells, 207
    relation to synapsis, 58
    relation to van der Waals-London forces, 85
  desmosomes (*maculae adhaerentes*):
    cell surface layer, chick, 157
    chick blastoderm, 60
    description and function, 60
    ectodermal cells, chick, 169
    enveloping layer cells, teleost, 93, 197, 200
    epithelial cells, 60, 92

Cell junctions (*cont.*):
  isolated cell membranes, 89
  neural fold cells, 163
  relation to intercellular material, 92
  relation to intercellular space, 60
  relation to reaggregation, 60
  relation to van der Waals-London forces, 87
  gap (20 Å):
    enveloping layer teleost, 192, 193, 197, 200
    fused cells, 80
    plasma membrane, 56, 58
  septate desmosomes:
    comparison with tight junctions, 62
    echinoderm invagination, 139
    electrical communication, 62
    function, 61
    invertebrates, comparison with desmosomes, 61
    relation to van der Waals-London forces, 87
  tight (*zonulae occludentes*, focal tight and zonal tight):
    cell contacts in primitive streak, 172, 173
    contact-inhibiting cell adhesions, 207
    ectodermal cells, chick, 169
    electrical communication, 58
    enveloping layer cells, teleost, 192, 193, 200
    epithelial cells, 56, 57, 80
    isolated cell membranes, 89
    neural fold cells, 163
    low electrical resistance, 173
    PASM-stained material, 93
    sites of calcium bridging, 80
  zonulae adhaerentes (or 100–200-Å gap):
    cell adhesion to glass substratum, 72, 85
    cell-to-cell adhesion, 78–80, 86, 87
    cells of surface layer, amphibian, 147
    deep cells, teleost, 191, 200
    ectodermal cells of chick, 169
    enveloping layer, teleost, 192–194, 197, 200
    epithelial cells, 58, 59
    intercellular material, 59, 62, 91, 92
    interdigitation, 62
    osmium tetroxide, 91, 92
    plasma membrane, 56, 57 (fig.), 58, 91
    primitive streak, chick, 172

Cell junctions (*cont.*):
  relation to hemoglobin and ferritin, 59, 93
  relation to van der Waals-London forces, 85, 92
Cell locomotion:
  axopodia, 207
  filopodia:
    activity in deep cells, teleost, 200, 204
    attachment to epithelium of epiblast, chick, 173
    chick retinal pigment cells, 122
    contracting and anchoring role, echinoderm, 143
    development in mesoderm cells, chick, 172
    formation from lobopodia, 199
    formation in deep cells, teleost, 190
    function during secondary phase of echinoderm invagination, 139–144
    mesenchyme migration in echinoderm, 12, 31–33, 142, 143, 204
    microtubule involvement, 144, 145 (ref.), 206
    organs of locmotion, 28, 203
    reaction to exogastrulation in echinoderm, 140
    relation to adhesiveness of echinoderm ectodermal cells, 144
    relation to cell motility, 204
    selective contact to ectoderm, 33
    tracheoles in insects, 204
  lobopodia:
    adhesive qualities, teleost epiboly, 199
    behavior in *Fundulus,* 34
    blastoderm, teleost, 28
    electron microscopic study late blastula and gastrula, teleost, 191
    formation in deep cells, teleost, 189
    late blastula, teleost, 187, 189
    organ of locomotion, 199, 203, 204
    teleost blastomeres, cluster movement, 129
  pseudopodia:
    activity on glass substratum, 83, 85
    amoeboid movement, 19
    effect of reversed acrasin gradient, 48, 49
    formation, 42
    relation to ruffled membrane, 22, 28, 33

Cell locomotion (*cont.*):
  role of pH gradient, amphibian gastrulation, 149
  ruffled membranes:
    cell orientation, 30, 85
    contact inhibition in fibroblasts, 22–26, 122
    epithelial sheet spreading, 127–129
    formation in culture, 204
    gliding movement, 20–22
    leading edge chick mesoderm cells, 172
    localized inhibition, 22–26
    marginal cells of enveloping layer, teleost, 190, 195, 196, 198
    mechanism of paralysis, 208
    motility of teleost gastrula cells in vitro, 188
    organ of locomotion, 203
    passive phenomena, 28
    relation to filopodia, 144
    teleost epiboly, 34, 200
    wound closure, 19
Cell markers:
  carbon, 8, 169, 185, 196 (*see also* Carbon marking)
  fluorescein, 19, 70, 208 (*see also* Fluorescein)
  natural, 104–107 (*see also* Reconstitution):
    melanin-protein granules, 87, 108, 119–122 (*see also* Chick embryonic cells, retinal pigment)
    nuclear size, 8
  radioactive isotopes, 8, 15, 108, 123, 171
    tritiated thymidine, 8, 10, 11, 15, 60, 64, 73, 108, 123, 171, 209 (*see also* Tritiated thymidine)
  vital dyes, 8, 11, 159, 171
Cell membrane (*see also* Cell adhesion):
  amoeboid movement, 19
  association with intercellular material, 63
  autonomous unit, 210–213
  cell adhesion, phospholipids, 56, 77, 85
  cholesterol, 88
  composition, 88
  decreased adhesiveness, 32
  deformability, 203, 204
    protein synthesis, 213
  electron microscopic study, 78, 80, 88, 166, 167 (*see also* Cell junctions)
  genetic influence, 1, 2, 211, 213

Subject Index   227

Cell membrane (cont.):
  methods of isolation, 88, 89, 201
  mucopolysaccharides, 85, 88, 94, 95, 146, 197 (see also Mucopolysaccharides)
  properties, 210–213
  proteins, 77, 88
  relation of differential gene function, 211
  self-replicating system, 213
  structural changes, 210, 211
  trilaminar, 56
  vitelline, 166, 167
Cell motility (see Cell locomotion)
Cell orientation (see also Cellular segregation):
  cellular slime molds (see also Cellular slime molds):
    contact following, 49
    radial, 48
    social phase, 47, 48
    solitary phase, 47
  colloidal micellae, 28, 31
  contact guidance, 28–31
  differential cellular adhesion, 30
  microexudates, 93
  substratum, 25, 26, 28, 30, 37, 70, 82, 83, 85, 93, 122–124, 152, 167
  theories, 30, 31
Cell sheet:
  contact inhibition, 128, 192, 207 (see also Contact inhibition)
  effect of mitosis on spreading and folding, 189
  effect of stearic acid, 30
  epithelial, 126–133, 197, 198, 204, 207 (see also Epithelial cell sheet)
  function in development, 5 (see also specific animal, gastrulation)
  palladium effect, 25
  reaction to wounding, 5 (see also Wound closure)
  spreading forces, 2, 4, 130, 133–135, 204
Cell streams, cellular slime molds:
  acrasin production, 50
  centripetal cell movement, 49
Cell surface (see also Cell adhesion):
  description, 56
  differential adhesiveness germ cells, 6
  effect of ion exchange, 175
  electron microscopic study, 56–63 (see also Cell junctions)
  genetic influence, 1, 2, 211, 213

Cell surface (cont.):
  modification during dissociation, 112, 115
  negative charge, 80, 81, 83
  permeability layer, 56
  rheological properties, 72
  role in contact inhibition, 22–26
  role of calcium, 80, 81, 90, 156
  unit membrane, 56–61, 78, 80, 86, 192 (see also Unit membrane)
Cell volume, neurulation, 162
Cellular exudate (see also Intercellular, material):
  cell orientation, 93
  formation of ground mats, 93
  future research, 97
  macromolecules, 30
  mucoproteins, 94, 95
  slime:
    production, 94, 95
    role of deoxyribonuclease, 95
  substratum, 93
Cellular segregation:
  chemotactic theory, 109
  differential cellular adhesiveness, 110–112, 115, 116, 118, 124
  discussion of theories, 112–124
  effect of pH change, 102
  internally segregating component theory, 112–118
  processes of cell cluster enlargement, 117–121
  qualitative factors, 117
  reconstitution, 100–109 (see also Reconstitution)
  theories, 109–124
  timing hypothesis, 110, 114
Cellular slime molds, 5, 36
  acrasin production in cell streams, 49 (see also Acrasin)
  agglutination, 77
  cell orientation in social phase, 47, 48
  center formation, 45–53
  chemotaxis, 40, 45–53 (see also Acrasin)
  contact following, 49
  development, 45
  founder cell, 47, 48
  leaders in study, 47
  mitogenetic rays, 45
  myxamoebae, 47
  relation of bacteria to social phase, 47

Subject Index

Cellular slime molds (*cont.*):
  secretion front formation theory, 51, 52
  social phase, 47, 48
  solitary phase, 47
  stippled aggregation, 49
  surface-specific antigenicity, 77
  time-lapse, 50, 51
Cellulose acetate, substratum, 25
Cellulose-ester polypore filter, substratum for spreading blastoderm, 167
Cement (see Intercellular, material)
Centrifugation:
  cell membrane isolation, 88
  measurement cell adhesive forces, 68
Centrifuge microscope, measurement distraction force, 72
Chaetopterus:
  artificial activation of eggs, 133
  development without cleavage, 133, 144
Chemotaxis:
  animal cells, 40–45
    Campanularia sperm, 43–45
    effect of various acids, 38, 40
    leukocytes, 42, 43
    lymphocytes, 42
    melanoblasts, newts, 40
  bracken fern sperm migration, 38
  cell aggregation, 37, 109
  cellular segregation, 109, 110, 112–119
  cellular slime molds, 45–53 (*see also* Acrasin)
  hydrodynamic flow, 38
  metabolites within aggregates, 109
  quantitative measure, 43
Chick embryo:
  desmosomes, 87
  elongation, 4
  germinal crescent, 5, 6
    incorporation tritiated thymidine, 8
  reaggregation patterns, 124
  reconstitution, 102
Chick embryonic cells:
  chondrogenic, 106, 107
  epithelium, 29–32
  heart:
    adhesion, 86, 87
    cellular segregation, 119–122
    fibroblasts in culture, 4, 23–29
    lens, microtubules, 205, 206
    liver, 73, 116

Chick embryonic cells (*cont.*):
  mesenphros, 104–106 (*see also* Reconstitution)
  pancreas, microtubules, 206
  retinal pigment cells:
    adhesion, 86, 87
    filopodial activity in aggregates, 122, 204
    locomotion, 143, 177
    segregation, 108, 112, 113, 117–123, 177
  wing bud, 104, 106
Chick gastrulation, 18, 166–177
  area opaca, 166–169
  basal lamina contacts, 173
  blastoderm, 189
    comparison with teleost, 189
    contact inhibition in spreading, 207
    culture, 66
    desmosomes, 60
    directional movement, 130
    marginal cell movement, 167, 168
    spreading, 167–179, 204
    substratum for hypoblast, 169, 170
  bottle cells (*see* Bottle cells, amphibian)
  cell coupling, 173, 174
  comparison with teleost gastrulation, 4, 169
  convergence, 4
  directional movement of hypoblast, 170
  endodermal adhesiveness, 176
  endoderm formation, 169–175
  epiblast, 172
    substratum for hypoblast, 169
  filopodia in mesoderm cell migration, 207
  heart cell cluster movement, 36, 122, 175–177
  hypoblast:
    definition, 169
    delamination, 4
    filopodial attachment mesoderm cells, 173
    in vitro, 169
    spreading, 169, 170
  mesoderm formation, 169–175, 199
  precardiac cell migration, 36, 175–177
  primitive streak formation, 4, 171
    electrical coupling, 175
    fine structure, 60, 172
  vitelline membrane:
    electron microscopic study, 166, 167
    substratum for blastoderm spreading, 166, 167, 168

Subject Index   229

Chordamesoderm:
  formation of neural tube, 159
  relation to amphibian epibolic spreading, 3, 4, 153
Close junctions (*see* Cell junctions)
Clypeaster invagination, 138
Coelomic pouch formation, 3
Colchicine:
  effect on microtubules, 207
  teleost epiboly, 189
Contact following, orientation in cell streams, 49
Contact guidance, 28–31
  amphibian gastrulation, 154
  cell orientation, 28, 31, 37, 38
  cell-to-cell adhesion, 93
  cellular slime mold, 46, 49
  chemotaxis, 42
  comparison with contact following, 49
  melanocytes, 31
Contact inhibition, 22–26, 32
  absence:
    blastoderm of *Fundulus,* 207, 208
    cellular segregation, 124
    deep cells of *Fundulus,* 190
  agar substratum, 25
  amphibian gastrulation, 153
  cancer cells, 22, 24, 25, 68, 207
  causes, 24, 25
  cell surface, 25, 26, 203
  cellular slime molds, 49
  chick blastoderm spreading, 207
  comparison with cellular segregation, 122–124
  contractility, 25
  effect of palladium, 25, 28
  electrical coupling:
  fibroblasts, 175, 208
  Fundulus gastrulation, 200
  enveloping layer in teleost, 207
  epithelial cell sheet, 128, 129, 207
  fibroblasts, 22–25, 122, 170, 208
  filopodia, chick mesoderm, 173
  glass substratum, 26
  hamster cells, 89
  hypoblast cells of chick, 170
  neural crest cells, 207
  ruffled membranes, 22–26, 122 (*see also* Cell locomotion, ruffled membranes)

Contact inhibition (*cont.*):
  tritiated thymidine, 123
Contact interspace, calcium binding, 80
Contractility:
  Amaroucium, caudal epidermis, 205
  ascidian tadpole, 144, 205
  bottle cells, 147–151, 164
  calcium binding, 81
  contact inhibition, 25
  endoderm, amphibian, 155
  filopodia:
    echinoderm, 12, 31–33, 139–144
    insect, 204
    teleost, 190, 199
  gliding movement, 22
  mechanism of neurulation, 163, 164
  microtubules (*see* Microtubules)
  teleost yolk surface layer, 183
Cortex:
  development in *Beroë,* 211, 212
  insect development, 212
  molluscan development, 212
  morphogenetic importance, 11
Coupling, electrical (*see also* Cell junctions):
  cancer cells, 208
  contact-inhibiting fibroblasts, 175, 208
  enveloping layer cells, teleost, 197, 200, 201
  ions, 173, 175, 197
  medullary plate, chick, 173
  mesoderm cell, chick, 175
  paralysis, ruffled membranes, 208
  relations to cell junctions, 208
  septate desmosomes, 62
  tight junctions, 58
Ctenophore, cortex in development, 211, 212
Cytoplasmic bridges, 63, 64
Cytoplasmic sheet as unit of activity, 133, 134

Deckschicht (*see* Enveloping layer)
Deep cells:
  teleost epiboly:
    absence of contact inhibition, 199, 207
    blastoderm spreading, 204
    description, 179, 181
    electron microscopic study, 190, 191, 200
    formation of lobopodia, 189
    migratory behavior, 12, 34, 189, 190, 199, 200
    relation to periblast, 189

Dendraster invagination, 136, 146 (ref.)
Deoxyribonucleic acid:
  cellular slime molds, 95
  exudate during cell dissociation, 91
  reduction in rate of synthesis during contact inhibition, 208, 209
Desmosomes (*see* Cell junctions)
Dictyostelium discoideum, 45, 46, 48–50, 52, 53, 77 (*see also* Cellular slime molds)
Differential cellular adhesion:
  early echinoderm gastrulation, 144
  germ cells, 6, 9
  Fundulus, 27, 28, 132, 200
  relation to cell orientation, 30
  theory of cellular segregation, 110–112, 115–118, 124
Directional movement:
  blastoderm in chick, 130
  deep cells in *Fundulus,* 201
  germ ring in teleost, 190
  hypoblast cells in chick, 170
  presumptive heart cell clusters, 130
  spermatozoa and spirochetes, 130
DNA (*see* Deoxyribonucleic acid)
Dulcioles, reaggregation, 16 (ref.)

Echinoderm eggs:
  advantages as research tool, 136
  electrostatic forces, 83
Echinoderm gastrulation, 4, 11, 12, 13, 32
  blastomeres, effect of calcium-free media, 19, 79, 90
  blastopore, 136
  blastula, 139
  comparison with amphibian gastrulation, 157
  comparison with neurula closure forces, 164
  comparison with teleost gastrulation, 199
  differential cellular adhesion, 144
  invagination, 11, 32, 136–145
    archenteron formation, 12, 33, 36, 133, 142, 143
    blastocoel formation, 36
    columnar cells, 138
    filopodial attachment, 143
    mesenchyme, primary, migration, 10, 15, 22, 31–33, 94, 104, 136–141, 144
    mesenchyme, secondary, migration, 8, 133, 138–144, 204
  septate desmosomes, 61

Ectoderm:
  behavior in culture, 102
  changes during neurulation, 161
  invagination to form brain, 4
  superficial, 8
  trunk neural crest cells, 8
Electrical coupling (*see* Coupling, electrical)
Electron microscopic studies (*see also* Cell junctions):
  amoeboid movement, 18
  amphibian invagination, 157, 169, 172
  blastoderm cell contact, teleost, 189
  bottle cells in amphibian, 148
  cell contacts, 56–63
  cell fusion, 64
  cytoplasmic bridges, 63, 64
  deep blastomeres in teleost, 190, 191
  enveloping layer in teleost, 192
  epithelial cells, 63
  erythrocytes, agglutination, 78
  glycoproteins, 62
  intercellular material, 62, 63, 92
  marginal cells, chick blastoderm, 167, 168
  membranes, adhering, 79, 80
  membranes, isolated L-cells, 88
  mesoderm formation, chick, 172
  microfilaments, 151, 157, 163, 203–207 (*see also* Microfilaments)
  microtubules, 144, 157, 163, 164, 172, 189, 203–207 (*see also* Microtubules)
  neural fold cells, 163
  surface coat, 147
  vitelline membrane, 166, 167
  yolk surface layer teleost, 182
Electrophoretic mobility, normal and transformed hamster cells, 89, 90
Electrostatic forces, cell adhesion, 83–86
Embryonic shield in teleost, 4, 36, 179–181, 190
Endoderm:
  culture with other germ layers, 102
  primordial germ cell organization, mouse, 7
  reconstitution, 104–106
  relation to chordamesoderm, 2
Enveloping layer (*see* Teleost epiboly)
Epiboly:
  amphibian, 3, 17, 133, 146–157
  chick, 4, 169–175
    proliferation center hypothesis, 170
  teleost, 34, 133, 179–201

Subject Index   231

Epithelial cells:
  behavior in vivo, 20
  desmosomes, 92
  extracellular filaments on intestinal cells, 62
  gliding movement in culture, 20–22
  reconstitution, 107
  tight junctions, 57, 80
  trilaminar membrane, 56
  zonula adhaerens, 59
Epithelial cell sheets:
  comparison with enveloping layer, teleost, 197, 198
  comparison with spreading blastoderm, teleost, 188
  contact inhibition, 128, 129, 207
  spreading in culture, 20, 22, 126–129, 133, 204
  spreading in wound healing, 127, 207 (*see also* Wound closure)
Epithelium:
  adhesive contact with substratum, 21
  composition of amphibian archenteron, 149
  desmosomes, 60
  germinal, 5
  response to wounding, 5 (*see also* Wound closure)
  selective adhesion, 103
  septate desmosomes in invertebrates, 61
Erythrocytes:
  adhesiveness during development, 27
  agglutination, 78, 86
  "ghosts," 88
Ester-linkage (*see* Cell adhesion)
Eye:
  microtubules, lens cells in vivo, 205, 206
  palisading of lens cells in culture, 205, 206

Feathers, microtubules in development, 206
Ferritin (*see* Cell junctions, *zonulae adhaerentes*)
Fetuin (*see* Cell adhesion, glass substratum)
Fibroblastic movement, in vitro and in vivo, 15, 31–33
Fibroblasts:
  adhesion to glass substratum, 29, 85
  adhesive and cohesive force, 70
  adhesive contacts with substratum, 21
  behavior in culture, 15, 19–26, 29–33, 70, 204

Fibroblasts (*cont.*):
  behavior in vivo, 15, 20
  chick heart:
    migration on various substrata, 4, 29–31
    ruffled membrane, localized inhibition, 23–26
  comparison with myxamoebae movement, 47
  contact inhibition:
    cell sheets, 128, 129
    ruffled membranes, 22–25, 122, 170
  contact relations with polyoma cells, 209
  electrical coupling, 175, 208
  gliding movement, 20
  microtubules, 206
  noncontact inhibition, 24, 45
  proteinacious exudate, 70
  ruffled membrane:
    comparison with deep cell filopodia, teleost, 200
    contact inhibition, 22–26, 122
    wound healing, 19
Filopodia (*see* Cell locomotion)
Flask cells (*see* Bottle cells)
Fluorescein:
  labeled antibodies, 70, 108, 124
  passage between coupled cells, 208
Focal close and tight junctions (*see* Cell junctions)
Forebrain, spreading activity, 152, 153
Founder cell (*see* Cellular slime molds)
Fundulus heteroclitus, 179–206
  cell adhesion during embryonic differentiation, 27, 28
  close and tight junctions, 58, 193
  electrical properties, 197, 200, 201
  filopodia, 34, 199–201, 204
  gastrulation (*see also* Teleost epiboly):
    behavior isolated blastoderm, 188
    blastomere cell cluster movement, 34, 129
    cell differentiation, 27, 28, 32, 132, 200
    comparison with amphibian gastrulation, 188
  desmosomes in enveloping layer cells, 200
  early cleavage, 200
  effect of colchicine, 189
  electrical coupling, 197, 201
  enveloping layer and periblast junction, 188, 195, 197, 199–201, 204
  epibolic forces, 199–201

Fundulus heteroclitus (cont.):
    locomotory behavior of deep cells, 34, 199–201, 204
    mucopolysaccharide layer, 197
    role of blastoderm in epiboly, 34, 170

Gastrulation (*see* specific animal)
Genes:
    activation of cell membrane changes, 211, 213
    control of morphogenetic movements, 213, 214
    effect on cell surface properties, 2, 211, 213
    transcription, 1, 213, 214
Germ cells:
    cytochemical identification, 7
    differential adhesion, 6
    glycogen, 6
    migration, 2
    primordial, 5–7
    selective adhesion, 55
Gliding movement:
    blue-green algae, 20
    contractile proteins, 22
    fibroblasts and epithelial cells in culture, 20, 21
    plant cells, 20
Glycoprotein:
    staining technique, 62
    yeast cell agglutination, 78
Gray crescent (*see* Amphibian gray crescent)
Ground mats, formation, 93
Gut:
    elongation in amphibian, 154
    expansion, 2
    formation, 5, 102, 103
    germ cells, 7
    primitive, 3

Haliclona (*see* Sponge cells):
    cellular segregation, 98
    reconstitution, 101
Hamster cells:
    electrophoretic mobility, 89, 90
    fine structure, plasma membrane, 57
    transformed, 89
Hansenula wingei, agglutination, 78 (*see also* Yeast cells)
Head-fold stage chick gastrulation, 175, 176

Head-process stage chick gastrulation, 176
Heart cells:
    chick, 4, 23–25, 29, 86, 87, 119–122 (*see also* Chick embryonic cells)
    mouse, cell junctions, 58
Heterotypic adhesions:
    cell aggregates, 11, 112
    cluster formation, 118
    comparison with isotypic adhesions, 73, 116
His hypothesis, 162
Hyaline plasma layer:
    echinoderm invagination, 139
        relation to columnar cells, 138
        relation to mesenchyme cells, 32
    production, 90, 91
Hydra, fine structure of epithelium, 67 (ref.)
Hydrogen bonding (*see* Cell adhesion)
Hydroides hexagonus, sperm and egg fusion, 64, 65 (fig.) (*see also* Annelid)
Hyla regilla:
    fine structure, 158 (ref.)
    microfilaments, 163

Imine bonds, cell adhesion, 76
Initiator cell (*see* Cellular slime molds)
Inosinic pyrophosphorylase, deficiency in polyma cells, 209
Insects:
    effect of cortex on development, 212
    filopodial distribution, tracheoles, 204
Intercellular:
    adhesion, definition, 74
    fluid, role in *zonula adhaerens*, 59
    material, 90–97
        calcium-bridge theory, 80
        desmosomes, 92
        electron microscopic studies, 62, 63, 92
        enzyme action, 91
        glycoproteins, 60
        mucopolysaccharides, 85, 88
        phospholipase C, 60
        properties, 63
        zonula adhaerens, 59
    space, desmosomes, 60
Internally segregating component (*see* Cellular segregation)
Intracellular deposits, desmosomes, 60, 61
Invagination, 3, 4, 11, 32 (*see also* specific animal, gastrulation)

Involution, 3, 133, 147, 149, 179 (*see also* specific animal, gastrulation)
Ion:
    exchange, surface activity of cells, 175
    flow, relation to electrical coupling, 173, 197
    pairing, mechanism cell adhesion, 77
Isoaffinity, salamander tissue, 103 (*see also* Reconstitution)
Isocitrate dehydrogenase, mouse embryo, 64
Isotypic adhesion:
    cluster formation, 118
    comparison with heterotypic adhesion, 73, 111, 116

Japanese medusa, sperm movement, 36

Kidney:
    reconstitution, 108
    tubules, induction, 94

Lanthanum staining:
    close junctions, 58
    intercellular material, 62, 92
Lecithin:
    cell adhesion, 85
    cell membrane, 88
Lens:
    cells, microtubules, 205, 206
    vesicle, brain formation, 4
Lesions, formation, 42
Leukocytes, 40, 69
    amoebocytes, 13
    amoeboid movement, 19, 32, 36
    negative chemotaxis, eosinophil, 42
    polymorphonuclear, 42, 43
Liver cells:
    chick:
        reaggregation, 73
        segregation in mixed aggregate, 116
    mouse, cell junctions, 58
Lymphocytes:
    amoebocytes, 13
    amoeboid movement, 32
    chemotaxis, 42
Lytechinus variegatus, invagination, 143

Macrophages, movement, 18, 19, 69
Maculae adhaerentes (*see* Cell junctions, desmosomes)

Magnetism:
    measurement amphibian gastrulation forces, 157
    measurement amphibian neurulation forces, 164
Mammals:
    gastrulation, 3
    neural tube formation, 159
Marginal cells:
    cell process as locomotor organs, 168
    enveloping layer, teleost epiboly, 188, 201
    epibolic spreading, chick, 167, 168
    spreading cell sheets, 126–129, 204
Melanin-protein granules (*see* Cell markers, natural)
Melanoblasts:
    migratory powers, 13
    negative chemotaxis, 40
Melanocytes:
    contact guidance, 31
    fibroblastic movement, 32
Melanophores, microtubules in pigment migration, 206
Mesoderm (*see also* specific animal, gastrulation):
    behavior in culture with other germ layers, 102
    evagination, 3
    extraembryonic, 130
Mespilia, invagination, 138
Metamorphosis, *Amaroucium,* 205
Methanamine, acid silver (PMAS):
    glycoproteins, 62
    intercellular material, 92, 93
Microciona prolifera:
    cellular segregation, 96
    reconstitution, 100, 101
Microfilaments:
    amphibian bottle cells, 151, 157
    cell shape changes, 204–207
    contractile epidermis of *Amaroucium,* 205
    formation, 207
    Hyla, 163
    neural groove cells, 163
    relation to microtubules, 207
    structure in cytoplasm, 203
Microspikes, intracytoplasmic structures in cell cultures, 205
Microtubules:
    absence in mesoderm filopodia, chick, 172
    amphibian bottle cells, 157, 164, 172

Microtubules (cont.):
  axopodia of *Actinosphaerium,* 207
  breakdown, 206–207
  cell shape changes, 204–206
  chick embryonic lens cells, 205, 206
  echinoderm filopodia, 144, 145 (ref.), 206
  feather development, 206
  melanophores, 206
  mitotic spindle, 206
  relation to microfilaments, 207
  structure in cytoplasm, 203
  teleost epiboly, 189
  Triturus alpestris, 163, 164
Microvilli:
  amphibian bottle cells, 148
  role in aggregation, 86
  teleost periblast, 191, 198
Millipore filter, substratum, 43
Mitogenetic rays (*see* Cellular slime molds)
Mitosis:
  absence in syncytial myofibrils, 209
  embryonic sheet behavior, 189
  neural tube formation, 162
  teleost epiboly, 169, 170, 189
Mitotic cycle, cell adhesion, 27
Mitotic inhibition, 209
  colchicine, 207
  contact inhibition, 208–209
  folic acid, 169
  spreading forces of chick area opaca, 170
  spreading forces telecost blastoderm, 170
Molecular complementarity hypothesis (*see* Cell adhesion)
Mouse embryonic cells:
  brain, 81
  electrostatic forces, 83
  heart, close junctions, 58
  kidney, 81
  liver, 58
  mesonephric reconstitution, 106, 107
  primordial germ, 6, 7
Mucopolysaccharides:
  cell membranes, 88
  extracellular material, 94
  intercellular cementing agent, 85, 88
  surface coat:
    amphibian, 146
    teleost, 197
Mucoproteins, cellular exudates, 94, 95
Myxamoebae (*see* Cellular slime molds)

N-acetyl-neuraminic acid (*see* Sialic acid)

Neoblasts, 13–15
Neoplastic cells (*see* Cancer cells)
Nerve fiber, regeneration, 104, 105
Neural crest:
  contact inhibition, 207
  derivatives, 7
  emigration of trunk cells, 8
  migration, 2, 7–11
Neural crest cells:
  accumulation, 10
  contact guidance, 31
  fibroblastic movement, 32
  migration, 36
  selective adhesion, 55
Neural retinal cells:
  aggregation factor, 95
  reaggregation, 73
  segregation in mixed aggregates, 116
Neuraminidase:
  cell deformability, 204
  effect on transformed hamster cells, 89
Neurulation, 159–167
  amphibian, 159–161
  differential contraction, 163, 164
  forces involved, 162–165
  microtubules and microfilaments, 163–165
  neural closure, measurement of forces, 164
  neural fold cells, fine structure, 163
  neural groove, microfilaments, 163
  neural plate:
    amphibian neural tube formation, 159–161
    spreading activity, 152, 153
  neural plate cells:
    behavior in culture, 164
    changes in shape during neurulation, 161
  neural tube, 7, 8, 36
    contact guidance, 31
    formation, 3, 159–161
    incorporation, tritiated thymidine, 10
    relation to Schwann sheath cells, 11
    role of water uptake, 162

Olfactory vesicle, formation, 4
Oogonia, formation, 5
Optic cup, 2, 4
Optic nerves, regeneration, 104, 105
Optic vesicles, formation, 4, 169
Orientation (*see* Cell orientation)
Oscillatoria, gliding movement, 20

Osmium:
    cell surface, 56
    intercellular material, 62, 92

Palladium (*see* Cell adhesion)
Paracentrotus lividus (*see* Sea urchin)
Paramecium, cell membrane, an autonomous unit, 211, 214 (ref.)
Passive cell movement:
    cell sheet spreading, 127, 128, 133
    hypoblast migration in chick, 169, 172
    nonmarginal cells in chick blastoderm, 168
    spreading, enveloping layer, teleost, 196, 198
Pigment cells (*see also* Chick embryonic cells):
    contact guidance, 31
    flank skin of newts, 36, 40–42
    negative chemotaxis in *Triturus*, 47
Planaria, regeneration, 14, 15
Plant cells:
    gliding movement, 20, 21
    microtubules in streaming, 206
Plasma clot as substratum, 29
Plasma membranes:
    close and tight junctions, 56–58, 85
    ectodermal cells during chick epiboly, 169
    interdigitation of enveloping layer cells in teleost, 197
    isolation of, 88–90
    junctional relations in chick epiboly, 172, 173
    role in cell fusion, 64
    structure, 56–58
    surface layer cells, amphibian, 147
Pluteus larva (*see* Sea urchin)
Polyoma transformed cells:
    inosinic pyrophosphorylase deficiency, 209
    relation to normal cells, 209
Polyoma virus, transformation of hamster cells, 89
Polysphondylium pallidum, 53 (*see also* Cellular slime molds)
Polysphondylium violaceum, 48–52 (*see also* Cellular slime molds)
Proliferation center hypothesis (*see* Epiboly)
Proteins:
    cell adhesion promotion, 83, 95, 96, 213
    cell membrane, 77, 88
    fibroblast exudate, 70
    genic control, synthesis, 213
    gliding movement, 22

Psammechinus miliaris (*see* Sea urchin)
Pseudocentrotus (*see* Sea urchin)

Rana pipiens:
    hybrid development and gene action, 156, 213
    neurula, 155
Reaggregation (*see* Cellular segregation)
Reaggregation kinetics, 72–74
Reconstitution, 100–109
    cartilage, 102
    epithelia, grafted, 103
    germ layers, amphibian, 102, 103
    kidney, 108
    mesonephric cells, 102
    mixed aggregates:
        ectoderm and endoderm, amphibian, 104, 106
        mesonephric and limb bud cells, chick, 104
        mouse mesonephric and chick chondrogenic cells, 106, 107
        nephric-tubule, 108
        retinal pigment, 108
    pronephric systems, amphibian, 102
    selective adhesion:
        isoaffinity and heteroaffinity, 103, 104
        regenerating nerve fiber, 104
    sponge, 100–102
Regeneration:
    amphibian, 13, 14
    contact inhibition, 207
    nerve fibers, 36, 104, 105 (fig.)
    Planaria, 14, 15
Reptile:
    neural tube formation, 159
    relation of yolk to cleavage, 3
Rhodnius prolixus, migration tracheoles, 204, 215 (ref.)

Saccloglossus, sperm and egg fusion, 64
Salamander larvae, selective adhesion, 103
Salmo irideus, epibolic forces, 186
Sand dollar (*see Dendraster*)
Sarcoma cells (*see* Cancer cells)
Sea urchin invagination (*see also* Echinoderm invagination):
    Lytechinus variegatus, 143
    Mespilia, 138
    Paracentrotus lividus, 91 (fig.), 138, 140, 143
    Pluteus larva, 32, 37 (fig.)

Sea urchin invagination (cont.):
   Psammechinus miliaris, 33
   Pseudocentrotus, 138
Selective adhesion (*see* Cell adhesion and Reconstitution)
Self-replicating system, 211, 212
Serranus atrarius, 212 (ref.)
Sialic acid:
   cell membrane, 89, 204
   relation to electrophoretic mobility, 90
Sorting out (*see* Cellular segregation)
Sphingomyelin, cell membrane, 88
Spirodon, behavior of sperm, 36
Sponge cells:
   cell cluster movement, 122, 129
   cellular segregation, 96, 100, 101
   filopodial activity in culture, 204
   locomotion, 143
   reaggregation, 96, 100, 101, 114
   reconstitution, 100, 101
   surface-specific antigenicity, 77
Staphylococcus albus, chemotaxis, 42
Substratum:
   agar, contact inhibition, 25
   cellular, 152
   glass:
      adhesive qualities, 70, 82–85, 122–124
      grooved, 30
   relation to cell flattening, 26
Syncytial myofibril, 63–66
   lack of mitotic activity, 209
Syncytial periblast, teleost cytoplasmic sheet, 134, 181

Tadpole, ascidian, contractile system, 144, 205
Teleost egg:
   cytoplasmic sheet movement, 134
   tools for epibolic studies, 179
Teleost embryo:
   adhesive and aggregative properties, 124
   cell cluster movements, 122
   desmosomes, 87
Teleost epiboly, 3, 4, 13, 18, 179–201
   blastoderm, 12, 133, 179–189
      behavior when isolated, 188
      cellular constituent, 187
      contractile tension of yolk surface layer, 183
      deep cells, 12, 34 (*see also* Deep cells)

Teleost epiboly (*cont.*):
      spreading activity, 179–181, 185–188, 207
      surface activities in situ, 189
   blastodisc, relation to yolk surface layer, 182
   blastomere:
      early cleavage, 200
      relation to periblast, 181
   "blastopore" closure, 179–181, 198
   blastula, 124, 179–181, 187, 188
   comparison with mesenchyme activity in echinoderm invagination, 199
   deep cells, 12, 143, 179, 181, 189, 190, 199
   embryonic shield, 4, 36
   formation, 179–181, 190, 199
   enveloping layer:
      cell-to-cell contacts, 192
      contact inhibition in spreading, 207
      deep cell lobopodia, 189, 190
      description, 179, 187
      electrical coupling, 197, 200
      elongation of cells during "blastopore" closure, 198
      forces in epiboly, 199–201
      junctions with periblast, 195, 196
      marginal cell activity, 188
      spreading activity, 197, 198
   germ ring:
      formation, 179–181, 190, 199
      formation of embryonic shield, 4, 36
   mechanism, theory, 197–201
   periblast:
      contact relations with enveloping layer, 195, 196
      formation, 181
      relation to blastomeres, 191, 192
      relation to yolk cytoplasmic layer, 182, 183, 196
      spreading, 201
      substratum for blastoderm, 183–185, 198
      substratum for deep cells, 189
      substratum for enveloping layer, 197, 198
   periblast epiboly, 196–199
   segmentation cavity:
      electrical coupling, 197
      formation, 189
   yolk surface layer, 182, 183, 196

Teleost neurulation, 4, 160
Template activity, cortical development, 213
Time-lapse cinemicrography:
  aggregating cells, 79
  amoeboid movement, 18
  cell streams, 50, 51
  chemotaxis, 43
  chick precardiac cell cluster migration, 175
  contact inhibition, 22–26
  echinoderm invagination, 32, 138, 139, 141, 144
  epithelial cell sheet spreading, 127, 128
  neurulation, 159
  pigment cells, mixed aggregate, 112, 113
  ruffled membrane activity, 21–25
  teleost epiboly, 12, 189, 195
Timing hypothesis, 110, 113, 115, 118, 119
Trilaminar membrane:
  epithelial cells, 56
  isolated cell membranes, 88, 89
Tritiated thymidine:
  chick mesoderm formation, 171
  liver cell movement, 15
  long-term cell marker, 8, 11
  migration of basal cells, 60
  neural crest migration, 10
  reaggregative studies, 73
  reconstitution studies, 108
  syncytial myofibrils, 209
  syncytium formation, 64
Triturus:
  differentiation of cell organelles, 200
  pigment cell migration, 47
Triturus alpestris:
  microtubules, 163, 164
  neural closure forces, 164
Triturus torosus, pregastrula movement, 16 (ref.)
Trout eggs, epibolic forces, 185, 187, 199–201

Unit membrane:
  cell adhesion, 78
  cell surface, 86
  close junctions, 58
  desmosomes, 60
  epithelial cell tight junctions, 80

Unit membrane (*cont.*):
  plasma membrane, 59
  septate desmosomes, 61
  structure, 56
  teleost enveloping layer cell interdigitation, 192
  tight junctions, 56–57

Van der Waals-London forces:
  desmosomes, 87
  relation to *zonula adhaerens,* 92
  role in cell adhesion, 84–86
Virus-transformed cells (*see* Polyoma virus)

Wound closure:
  cell movement, 32
  cell sheet, 5
  contact inhibition, 207
  epithelial cell sheet spreading, 127
  fibroblasts, 19
  yolk surface layer, teleost, 182

Xenopus:
  cortical differentiation, gray crescent, 211, 212
  microfilaments, 163
Xenopus laevis, cortical inheritance, 214 (ref.)

Yeast cells, agglutination, 78
Yolk:
  relation to amphibian cleavage, 3
  teleost epiboly, 134, 179–181, 183, 198
Yolk cytoplasmic layer, teleost, 133, 196, 199
Yolk plug, relation to amphibian bottle cell formation, 151
Yolk sac:
  in culture, 5
  endoderm in mutant mouse, 7
  formation in chick, 4
  formation in teleost, 179
Yolk-surface-layer hypothesis, contractile tension, 182–185
Yolk syncytium, 66 (*see also* Teleost epiboly)

Zebra fish, surface layer tension, 182